CW00517884

Spatial Reasoning in the Early Years

Principles, Assertions, and Speculations

Brent Davis
and the Spatial Reasoning Study Group

Routledge
Taylor & Francis Group

NEW YORK AND LONDON

First published 2015
by Routledge
711 Third Avenue, New York, NY 10017

and by Routledge
2 Park Square, Milton Park, Abingdon, Oxon, OX14 4RN

Routledge is an imprint of the Taylor & Francis Group, an informa business

© 2015 Taylor & Francis

The right of the editor to be identified as the author of the editorial material, and of the authors for their individual chapters, has been asserted in accordance with sections 77 and 78 of the Copyright, Designs and Patents Act 1988.

All rights reserved. No part of this book may be reprinted or reproduced or utilised in any form or by any electronic, mechanical, or other means, now known or hereafter invented, including photocopying and recording, or in any information storage or retrieval system, without permission in writing from the publishers.

Trademark notice: Product or corporate names may be trademarks or registered trademarks, and are used only for identification and explanation without intent to infringe.

Library of Congress Cataloging in Publication Data
A catalog record for this book has been requested.

ISBN: 978-1-138-72903-6 (hbk)
ISBN: 978-1-138-79204-3(pbk)
ISBN: 978-1-315-76237-1 (ebk)

Typeset in Palatino and Myriad Pro
by Brent Davis

Spatial Reasoning in the Early Years

Over the past several years, "spatial reasoning" has gained renewed prominence among mathematics educators, as spatial skills are proving to be not just essential to mathematical understanding but also strong predictors of future success beyond the classroom in fields such as science, technology, and engineering. By exploring both primary and emergent dimensions, *Spatial Reasoning in the Early Years* helps define the concept of spatial reasoning and provides compelling evidence of the need for a clear focus within early education specifically. The authors review the research, look across current theories, and investigate implications for contemporary school mathematics pedagogy as they identify areas of inquiry necessary to bring a stronger spatial reasoning emphasis into the classroom.

The book contains many classroom- or workshop-based vignettes, highlighting the complexity of spatial reasoning in educational practice, providing an in-depth analysis of spatial reasoning as it applies to classroom practice, and offering new ways of framing lessons to help young students hone their spatial reasoning abilities. The book concludes with a forward-looking agenda that contributes to developing a greater understanding of the role spatial reasoning plays in educational contexts and beyond. Supported by plentiful visual representations, *Spatial Reasoning in the Early Years* skillfully integrates the conceptual and the concrete, making this text a dynamic and accessible resource.

Brent Davis is Professor and Distinguished Research Chair of the Werklund School of Education at the University of Calgary, Alberta, Canada.

The **Spatial Reasoning Study Group** is a transdisciplinary team of researchers from across North America with its hub at the University of Calgary. Its members work in and across mathematics education, mathematics, psychology, curriculum studies, and cognitive science. The group currently includes Catherine D. Bruce, Beverly Caswell, Lissa D'Amour, Brent Davis, Michelle Drefs, Krista Francis, David Hallowell, Zachary Hawes, Donna Kotsopoulos, Lynn McGarvey, Joan Moss, Yukari Okamoto, Paulino Preciado, Nathalie Sinclair, Diane Tepylo, Jennifer S. Thom, and Walter Whiteley.

Contents

Contributors

Catherine D. Bruce is Associate Professor in the School of Education at Trent University, Ontario, Canada. She studies children's learning of mathematics including fractions, algebra and the development of spatial reasoning.

Beverly Caswell is the Director of the Robertson Program for Inquiry-based Teaching in Science and Mathematics at the University of Toronto, Ontario, Canada. She works with educators to design learning environments that promote deeper understanding of concepts and equity.

Lissa D'Amour is Assistant Professor in the Werklund School of Education, University of Calgary, Alberta, Canada. She studies the interplay of anxiety and attachment security in dynamic teacher–learner systems.

Brent Davis is Professor and Distinguished Research Chair in Mathematics Education in the Werklund School of Education, University of Calgary, Alberta, Canada. He researches the educational relevance of the complexity sciences, focusing in particular on teachers' disciplinary knowledge of mathematics.

Michelle Drefs is Assistant Professor and Director of Training for the School and Applied Child Psychology program in the Werklund School of Education, University of Calgary, Alberta, Canada. Her research interest is in the assessment of early numerical competencies.

Krista Francis is Assistant Professor and Director of the Imperial Oil Science Technology Engineering and Mathematics (IOSTEM) initiative with the Werklund School of Education, University of Calgary, Alberta, Canada. Her research focus is K–12 STEM teacher professional learning.

David Hallowell is a doctoral candidate in Education at the University of California Santa Barbara. He holds M.A. degrees in both Education and Philosophy. His research focuses on young children's spatial reasoning.

Zachary Hawes is a Research Officer at the Dr. Eric Jackman Institute of Child Study, a research center and laboratory school of the University of Toronto, Ontario, Canada. His research examines the development of children's spatial and numerical cognition, with a focus on the effects of classroom-based interventions.

Donna Kotsopoulos is Associate Professor of Education and the director of the Mathematical Brains Laboratory at Wilfrid Laurier University, Ontario, Canada. Her research focus is on the human sciences. She studies mathematical cognition and learning across the lifespan.

Lynn McGarvey is a Professor of Mathematics Education at the University of Alberta, Alberta, Canada. Her research focuses on mathematical reasoning of young children with emphasis on the topics of algebraic thinking and spatial reasoning.

Joan Moss is Associate Professor in Department of Applied Psychology and Human Development at the University of Toronto, Ontario, Canada. Her research has focused on young students' development of rational number, early algebra, and spatial reasoning.

Yukari Okamoto is Professor of Education at the University of California Santa Barbara. She specializes in children's numerical, spatial, and biological reasoning. She received her Ph.D. from Stanford University and was a Spencer Foundation Postdoctoral Fellow.

Paulino Preciado is Assistant Professor and Director of the Math Minds Initiative in the Werklund School of Education, University of Calgary, Alberta, Canada. His research focuses on the collaborative design of tasks and lessons aimed at promoting deep mathematical understanding through intellectual engagement.

Nathalie Sinclair is Professor in the Faculty of Education and a Canada Research Chair in Tangible Mathematics Learning at Simon Fraser University, British Columbia, Canada. She is the author co-author of *Mathematics and the Body: Material Entanglements in the Classroom,* among other books.

Diane Tepylo is mathematics teacher and a Ph.D. candidate in the Department of Applied Psychology and Human Development at the Institute for Studies in Education, University of Toronto, Ontario, Canada. Her research focuses the role of spatial reasoning in mathematics and teacher-centered professional learning.

Jennifer S. Thom is Associate Professor at the University of Victoria British Columbia, Canada. Her research focuses on mathematics in the early years as well as the embodied and collective nature of mathematics and mathematical understanding.

Walter Whiteley is Professor of Mathematics and Statistics. He is a member of the Graduate Programs in Mathematics and a member of Education, in Computer Science and Interdisciplinary Studies, York University, Ontario, Canada. He is a researcher in discrete applied geometry (in engineering, sciences, and mathematics), and a geometry educator for future teachers and for future mathematicians.

Acknowledgments

The Spatial Reasoning Study Group (SRSG) is a transdisciplinary team. Its members work in and across mathematics education, mathematics, psychology, curriculum studies, and cognitive science. The group currently includes Catherine D. Bruce, Beverly Caswell, Lissa D'Amour, Brent Davis, Michelle Drefs, Krista Francis, David Hallowell, Zachary Hawes, Steven Khan, Donna Kotsopoulos, Lynn McGarvey, Joan Moss, Yukari Okamoto, Paulino Preciado, Nathalie Sinclair, Diane Tepylo, Jennifer S. Thom, and Walter Whiteley.

The SRSG has met regularly since 2012. This book is its first major print publication, following up on a series of presentations and symposia at major conferences (e.g., Bruce et al., 2013; Sinclair & Bruce, 2014). Like those conference engagements, this text is better described as a "co-authored publication" than a "joint presentation" or an "edited collection." Unfortunately, we are unable to signal the extent of members' contributitions to each part of this book, but we are keen to underscore that the listed authors for the chapters are convenient fictions. They indicate who took major responsibilities, but they do not reflect the depths of the shared thinking and collaborative efforts.

The work of the SRSG is made possible through the support of the Imperial Oil Science, Technology, Engineering, and Mathematics Initiative. IOSTEM is housed in the Werklund School of Education at the University of Calgary.

The cover image is from an original photograph provided by Steven Khan. We would like to express our appreciation to Jonna La Joy, Meghna Soni, and Yu Zhang at the University of California Santa Barbara and to the Math 4000 project group at York University for their helpful suggestions.

Section 1

What is spatial reasoning and why should we care?

SECTION COORDINATOR: YUKARI OKAMOTO

Chapter 1: What is spatial reasoning?

Spatial reasoning has always been a vital capacity for human action and thought, but has not always been identified or supported in schooling. In recent years we have noticed its role in maneuvering through and managing one's world much more. We survey its history and the recent recognition of its growing importance – within mathematics, across other disciplines, and in life beyond the school.

Chapter 2: The development of spatial reasoning in young children

Drawing primarily from the psychological literature, this chapter describes what we currently know about young children's development of spatial reasoning. We provide an overview of spatial reasoning as a multifaceted construct through concrete examples. This chapter provides a context for the chapters to follow.

Chapter 3: Developing spatial thinking

Once believed to be a fixed trait, there is now widespread evidence that spatial reasoning is malleable and can be improved in people of all ages. In this chapter, we first discuss the relationship between spatial reasoning and mathematics and then present a variety of spatial training approaches shown effective in not only supporting children's spatial reasoning but mathematics performance as well.

1

What is spatial reasoning?

WALTER WHITELEY, NATHALIE SINCLAIR, BRENT DAVIS

> ── In brief ... ──
>
> Spatial reasoning has always been a vital capacity for human action and thought, but has not always been identified or supported in schooling. In recent years we have noticed its role in maneuvering through and managing one's world much more. We survey its history and the recent recognition of its growing importance – within mathematics, across other disciplines, and in life beyond the school.

Emerging emphases on spatiality

In recent years, large-scale and highly influential organizations, such as the National Council of Teachers of Mathematics (2010) and the National Research Council (2006), have advocated for a more "spatial" approach to the teaching and learning of K–12 curricula.

This push to emphasize spatial reasoning signifies a shift in educational values and long-term educational objectives. These transitions are most frequently associated with the growing need for expertise in science, technology, engineering, and mathematics – the STEM disciplines. An important element in increasing STEM participation and success, it is widely asserted (e.g., Newcombe, 2010; Uttal et al., 2013), is to increase the education and development of spatial thinking.

The suggestion is not unwarranted. Extensive research has shown that spatial reasoning ability and success in STEM domains are strongly correlated (e.g., Wai, Lubinski, & Benbow, 2009). High school students who demonstrate strong spatial skills are more likely to enjoy, enter, and succeed in STEM disiciplines. Well-honed spatial skills are also associated with innovation by practitioners within these fields (Kell et al., 2013). The relation between spatial thinking and mathematics is pronounced; people with strong spatial skills are generally successful at mathematics (Wai et al., 2009; Mix & Cheng, 2012). These findings suggest that developing spatial thinking skills may support interest and achievement in STEM.[1]

Our particular focus in this book is on the development of spatial reasoning in early years (birth to Grade 3) mathematics, but it is important to note that spatial reasoning plays a vital role across all grades and within most, if not all, academic subjects. In this opening chapter, we situate the early years focus within the K–16 experience. This chapter also serves to frame our broad, collective intention to survey the available literature – historical and contemporary, academic and professional – with an eye toward providing a broad-based-but-useful introduction to the array of perspectives, the multiple fields of research, the spectrum of interpretations, and range of practices and children's activities that are collected under the banner of spatial reasoning.

For us, this project demands close and sustained attention to both historical circumstances and emergent situations. Paying attention to the past helps us to understand how varied constructs of spatial reasoning have contributed to (or have not contributed to) current conceptions of mathematics as a network of mathematical science disciplines and popular enactments of school mathematics. Paying attention to the present compels us to be mindful of what it means to be educated in a rapidly changing world. And so, while we aim to speak to the realities of entrenched school mathematics, we also look toward a model of school mathematics that is fitted to emergent personal, social, cultural, and ecological circumstances.

Our spatial reasoning project

This book is a tightly integrated discussion, not an edited collection of papers. While principal authorship for each chapter has been attributed to specific individuals (based on weight of contribution), every member of our Spatial Reasoning Study Group took part in structuring, critiquing, and integrating every chapter.

That detail is of particular relevance because our project is deliberately transdisciplinary. Our group comprises individuals with expertise in psychology, mathematics and mathematics education. Given that diversity of specialization, one of our first shared realizations when we began to discuss the research on spatial reasoning was that core constructs are often treated in radically different ways across domains, in spite of highly similar vocabularies. (See Chapter 9.) This point is particularly obvious around diverse meanings, varied methodologies and key examples associated with the phrase "spatial reasoning."

For example, there is considerable debate on the relationships among "visualization," "visual-spatial reasoning," and "spatial reasoning." Some commentators use these terms interchangeably; others offer subtle distinctions; some argue they must not be conflated. Another topic that is engaged in various ways is the role of language in the development of spatial reasoning abilities, with extremes of opinion being, at one pole, they are inextricably intertwined and, at the other, spatial reasoning can develop relatively independently from language.

We do not aim to resolve these debates. On the contrary – and as undertaken mainly in Chapters 2 and 3 and returned to in Chapter 9 – a main strategy in the book is to preserve differences in perspective in order to bring them into productive conversation. That is, our focus is not an unambiguous definition, but a working knowledge that will support richer understandings of how children engage with/in their worlds. That said, our project is to contribute to a more nuanced discussion of the phenomenon of spatial reasoning. We thus embrace work that tightly links spatiality to the visual – as exemplified, for example, in Tahta's (1989) work. He proposed the following three "powers" involved in working with space:

- *imagining*, which involves *seeing what is said*;
- *construing*, which involves *seeing what is drawn* or *saying what is seen*; and
- *figuring*, which involves *drawing what is seen*.

While Tahta did not use the phrase "spatial reasoning," for us his three powers illumine the developmental interplay of practices, highlighting in particular the movement in reasoning to and from language. These practices provide a powerful basis on which to organize a curriculum and a useful starting point for articulating our own working understanding of spatial reasoning.

Through our varied and collected experiences within mathematics and learning spatial reasoning, we have extended Tahta's and others' lists to include a range of dynamic processes that we see as characterizing spatial reasoning – and that can, but do not necessarily, involve concurrent work with language. Our preliminary list of verbs follows:

- locating
- orienting
- decomposing/recomposing
- shifting dimensions
- balancing
- diagramming
- symmetrizing
- navigating
- transforming
- comparing
- scaling
- sensing
- visualizing.

We are attempting to be suggestive here, not exhaustive. To that end, we revisit this list – critiquing, extending, and re-organizing it – in the closing chapter.

We see both theoretical and practical value in the exercise of identifying and organizing such aspects. On the level of theory, these verbs compel us to be clearer on what we imagine ourselves to be discussing. And in the realm of practice, for example, one way for classroom teachers to assess whether a lesson is tapping into spatial reasoning is to determine whether some of these dynamic processes are being targeted or called into action.

In somewhat different terms, this book is founded on the assumption that spatial reasoning is not an isolatable competence, nor one that can be parsed into discrete sub-skills. Rather, it is an ever-evolving potential that arises within the complex interplay of many aspects. This lens on spatial reasoning and the spatial dimensions of knowing reminds us that what it means to "do mathematics" cannot be dissociated from everything else that humans study and do – a point that we regard as a resurgence of a waning sensibility. For example, until the beginning of the 20th century, people who practiced mathematics also typically engaged in science. In particular, spatial reasoning figured prominently in thinking about physical properties and laws: Euclid wrote a book on Optics; Pythagoras and Archimedes worked in a world in which geometric reasoning was more trusted than arithmetic. Studies in the medieval university were organized in large part around the Quadrivium, comprising disciplines that are both number-focused (i.e., Arithmetic and Music) and space-focused (i.e., geometry and astronomy – see Figure 1.1). The applied spatial domain of Astronomy was of particular relevance, foundational as it was to the work of traders and others needing to locate, navigate, and survey (see Henderson & Taimina, 2005, "Introduction"). As recently as a century ago, it was common for such prominent mathematicians as Hilbert, Poincaré and Noether to be simultaneously occupied with the physical and applied sciences, especially Physics and Engineering. In other words, the dissociation of spatial reasoning from the practices of mathematics, both in formal educational settings and in popular conception, is recent.

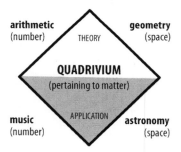

Figure 1.1: The Quadrivium – which, along with the mind-focused Trivium of Logic, Grammar, and Rhetoric – constituted the core of study in the medieval university.

This point might be illustrated by looking at the treatment of geometry in school mathematics. As detailed in Chapter 4, a principal motivation for the first government-funded public schools was to fill needs for workers with particular skills created by industrialization, urbanization, and capitalism. A consequent priority in public schools of the era was basic numeracy, which was tethered to a diminished status of geometry in formal schooling. Much in contrast to mathematical study of earlier centuries, in which the sciences and geometry were obvious and rich locations for spatial reasoning, the mathematics of the Industrial-era school was stripped down to the bones of useful arithmetic and formulaic algebra. Not only was most geometric content removed, the elements of arithmetic and algebra that remained were formatted with minimal demands for visualization and spatial reasoning. That trend was actually amplified in the mid-20th century, when school mathematics saw an even further erosion of topics in geometry (Whiteley, 1999). Combined, such happenings have contributed immensely to the lacuna in opportunities to engage in spatial reasoning in the context of school mathematics.

On this point, it is not without irony that school mathematics, as experienced by most citizens of the modern world, has come to be mainly about "figuring things out." That expression traces back to the 15th century, originally having to do with "picturing in the mind." However, by the 19th century, the meaning of the phrase had shifted almost completely to "working out a sum," and the project of public schooling was doubtless complicit in the shift. This demonstrated power of formal education to format popular opinion, of course, means that the institution might also contribute in transforming the situation in a more productive direction – and there are indeed signs that things are changing. For instance, the recent publication of mathematics undergraduate textbooks with titles such as *Visual Group Theory* (Carter, 2009), *Visual Complex Analysis* (Needham, 2004), and *Visual Geometry and Topology* (Fomenko, 2011), as well as titles such as *Vision in Elementary Mathematics* (Sawyer, 2003), provide some evidence of light making it through the cracks of the algebraic wall of the secondary, post-secondary (and now, in some places, also the pre-secondary) mathematics curriculum.

But the return of the spatial – as well as the geometric – has been a long time coming. Whiteley (1999) chronicled its initial demise, which stretches back at least to the anxieties in mathematics arising out of the possibility of non-Euclidean geometries, and to the so-called "foundations crisis," which laid the ground for the Bourbaki-style mathematics in which the alphanumeric triumphed over the visual, spatial, and dynamic modes of reasoning. The fact that mathematicians over the past 40 years, from Jean Dieudonné and René Thom to Sir Michael Atiyah,[2] have had to assert the importance of visual thinking to the development and understanding of mathematics, however, is a sign that this viewpoint is not widely held.

Against these backdrops, one of our intentions with this book is to mount a multi-plied – that is, many-layered – argument in favor of "radical spatialization" of the curriculum. Such is the particular focus of Chapters 6 through 9, in which we frame educational possibilities through a series of interpreted vignettes. Our discussion culminates in the assertion that emergent insights into spatial reasoning should present major challenges to the structures and emphases of contemporary school mathematics. Despite the profound natures of these challenges, there is a risk of a trivialized uptake – of, for example, taking up principles only insofar as they might support the development of entrenched topics.

Put differently, we believe that a serious consideration of emergent research insights into spatial reasoning not only compels a closer alignment between school mathematics and the field of mathematics. Spatial reasoning also provides one means to pull the two enterprises together, oriented by the conviction that they should share some core features including ways of problem solving. Briefly, since mathematicians depend on visual forms, as Atiyah (2002) wrote, so should learners. Hawkins (2000) argued that the major developments in mathematics have come about through deep connections between the quantitative and the qualitative (i.e., between number and shape, arithmetic and geometry, or computational and spatial). The Cartesian coordinate system is a powerful example, demonstrating that the combined powers of these two forms of reasoning can produce insight and understanding.[3] The corollary is that strength in both types of representation, and reasoning that combines work from both is something that students should be developing.

The development of spatial reasoning is, of course, already well underway by the time a child reaches the classroom door. Humans enter the world after 9 months in tight quarters, where they begin their learning about orienting, motion, distances, and other aspects of occupying space. By birth, then, children are already spatialized beings on a rapid trajectory of brain development as they move through, explore control of their bodies within, and manipulate aspects of their world. As they learn to roll, to crawl, and then to walk, they are learning balance in space – and, correspondingly, to "balance space." (Children just learning to walk, for example, often grasp or carry objects in *both* hands, but rarely in just one, indicating a sort of symmetry-making in their actions.) They imitate adult motions, often demanding mental rotations (e.g., infant sign language involves right hand following right hand). Children reason out what they can reach and access, and they soon start to connect these reasonings to their encounters with conventional images in books, on screens, in toys, and so on. Alongside these activities, they learn language for spatial relations and notice what adults point to and value. As with mathematics, their spatial reasoning is fundamentally associated with their mobility and their seeing.

Through selective engagement and a unique repertoire of explorations in space, each child develops a spatial sense that is unavoidably idiosyncratic. Idiosyncrasies are, of course, mitigated through common experiences, shared vocabularies, and more structured interventions such as preschool programs and activities – including, notably, the deeply spatial activities of Froebel's original, mid-1800s Kindergarten (Brosterman, 1997). Most children show interest in plane symmetry (a visual experience) and demonstrate abilities to create symmetric designs. As a result, they arrive at Kindergarten and Grade 1 having made robust connections among knowings that researchers identify as logical and spatial reasoning skills.

From a pedagogical point of view, our discussion rests on two facts: firstly, children come to school with a tremendous repertoire of informal spatial understandings that can and should be developed (as we will show in Chapter 2). Secondly, spatial reasoning supports mathematical understanding and problem solving (as will be shown in Chapter 3). Indeed, as Tahta (1980) noted, even the most abstract mathematical objects are described using geometric metaphors, citing the example of Dedekind, for whom a set was a bag, and Cantor, for whom infinity was an abyss. Turning to school mathematics, Tahta wrote: "At a more important elementary level, the failure of so many to handle numbers confidently may be due to the fact that they do not have any mental picture corresponding to the numerals on which they are required to operate" (p. 3).

More broadly, Newcombe (2014) describes an entwined development of spatial and numerical development. These are unavoidably connected in young children and language, but school culture and curriculum conventions conspire to separate them. The separation is relatively rare outside of school, however, as might be underscored through three observations:

- From a more economic point of view, a growing body of research points to the importance of spatial reasoning in a large range of jobs including computer-graphics, medical imagery, and engineering – among many others (cf., Wai et al., 2009).

- From a more pragmatic, everyday view, as our society grows ever more information dense, the interfaces used to access and manipulate that information are shifting away from the alphanumeric toward the spatial.

- From inside the world of children today, through focused and prolonged engagements with video games and other highly spatial interfaces, society might inadvertently be educating highly sophisticated spatial reasoning abilities.

For us, such observations point to the need for a more deliberate and better-developed approach to the educational support for spatial reasoning capacities.

The various chapters in this book will provide ample evidence for and examples of these claims. One focus of this chapter is to attend specifically to the second one, that spatial reasoning supports mathematical understanding and problem solving. We do this in part because of our firm belief that psychological and educational studies in mathematics education can be greatly improved by paying close attention to spatial reasoning within the disciplines of the mathematical sciences, including its changing historical and cultural values.

A key point in all of this is the relatively recent realizations that, just as spatial reasoning can be learned, it can atrophy, and it can be developed at any age. As Hoffman (2000a) put it, "we create what we see" – with the corollary that we can change (or re-create) what we notice and what we ignore. This is true even more broadly for spatial reasoning. A vital insight in this regard is that spatial reasoning can also fail to be developed, be underdeveloped, or can remain disconnected from other aspects of thought, themes that emerge repeatedly across the chapters that follow.

What is (not) spatial reasoning?

Above we offered a preliminary list of aspects of spatial reasoning. That list was derived in large part from others' research over the past three decades. For some time now, mathematics educators have focused on the importance of visual thinking and of visualization in mathematics education (e.g., Dreyfus, 1994; Presmeg, 2008). The more recent attention to spatial reasoning seeks both to address broader forms of thinking that may not be strictly visual in nature and to legitimize these forms of spatial thinking as being truly epistemic.

In Chapter 2 we offer a more detailed account of the emergence of the construct of spatial reasoning, drawing in particular on Uttal and colleagues' (2013) meta-analysis of spatial practices and tests from the social sciences (including psychology and education). Their typology captures some key distinctions in fine-grained practices and representations that are testable and are incorporated within the literature on spatial reasoning. However, as we explore in greater detail in Chapter 3, instances of children's mathematical spatial reasoning can rarely be identified as a singular "type"; rather, their reasoning typically involves multiple steps that move between and span categories. Further, it is not clear that carving up spatial reasoning into a collection of mutually exclusive types helps address the question of when and why spatial reasoning would be desirable when working mathematically. For this, it would seem important to focus more specifically on the *reasoning* part of spatial reasoning. Spatial *reasoning* must be more than spatial *awareness* – especially when it is recalled, as Tahta (1980) pointed out, that babies learn to stand on their feet and walk.

Addressing the word *reasoning* is made challenging in part because of the linguistic forms with which reasoning is usually associated. As many commentators (e.g., Ong, 1982; Donaldson, 1986) have noted, reasoning tends to be associated with words[4] – and, more specifically, with the stability and frozen linearity of written text. The associations run deep in most European languages, evidenced by the roots of the word *logic*, which is derived from the ancient Greek *logos*, "reason, idea, word." Spatial reasoning is a specific area of non-verbal reasoning, with a suite of powerful tools. Therefore, when we speak of spatial reasoning, we are speaking of explanations of some sort (of why or when or how), and so we first need to understand the various non-verbal ways in which reasoning might (and should) occur (including through gestures, diagrams and mental imagery) and also need to value, in the classroom, these forms of reasoning.

A different approach to the question of describing the phenomenon of interest can be found in the work of Gattegno (1963), who sought to capture the distinctive difference between geometry and algebra. In particular, as Tahta (1980) explained, Gattegno proposed that "geometry is an awareness of imagery" (p. 6) while algebra is the formalization of such awareness. Oriented by this assertion, Tahta argued that the premature shift to algebra, which is prevalent in school mathematics, can have devastating consequences for learners by robbing them of something to act upon. Tahta's definition places imagery in a central role, suggesting that geometry is about working with – summoning, drawing, transforming, imagining, becoming aware of – imagery, which highlights its important connection to spatial reasoning. In his work with the Nicolet films (which were precursors to dynamic geometry in terms of telling stories about the dynamics of shapes), Tahta worked on the awareness of imagery by inviting learners to watch, talk about, and imagine constrained movements of shapes.

For us, this kind of work emphasizes the importance in mathematics of variance and invariance. This seems to be an important point in terms of how we might productively think about spatial reasoning in mathematics education, namely, that there is often a mathematical goal that circumscribes instances of spatial reasoning, which is to investigate families of configurations in order to identify regularities, breaches, and invariances. In this sense, spatial reasoning can be thought of as fundamentally mobile – that is, open to variation in shape and in movement over time. Within the mathematics education literature, this thinking has been collective under the notion of "embodiment," which is the particular focus of Chapter 5.

It is also important to recognize what we do not include in spatial reasoning, as well as the inclusiveness of our use of the term. There are a number of visual reasoning practices that we are not including under spatial reasoning. Studies associated with the term "diagrammatic reasoning" (and the regular workshops of the community developed around this interest), as well as activities captured in the "Visual Literacy" community include many human practices, often with developed conventions and abstractions that blend actions and experiences that originated in vision and have been cognitively blended within practices of various communities (SIGGRAPH, 2002). Thus, for example, working with diagrams can be an important part of spatial reasoning, but not all work with diagrams uses reasoning within the spatial metaphors. Within this book, spatial reasoning will focus on examples that include spatial information (spatial locations, spatial adjacency, symmetry, transformations) and reasoning with representations that provide access to that spatial information.

It is important to underscore that spatial reasoning is learned. Pictures and diagrams are not, contrary to what is assumed in many textbooks, transparent conveyors of meaning; they have a learned grammar of their own. Sacks (1993), in his essay on the experience of recovering sight after a lifetime of blindness, reminded us:

> When we open our eyes each morning, it is upon a world we have spent a lifetime learning to see. We are not given the world: we make our world through incessant experience, categorization, memory, reconnection. (p. 71)

This is true for children as well as adults. It is also true for all areas we include under the umbrella of spatial reasoning. This world is three dimensional (3D), and 3D space is the native world of children. The 3D world is what our brains develop to handle with the minimal cognitive load (Hoffman, 2000b). Spatial reasoning can be practiced with limited or no use of the eyes – with the hands, with the moving body and gestures, and in cognition with kinesthetic sense and mental rotation. We have learned about space, and reason about space, from touch and movement, even in the womb, as well as from the eyes. Sometimes, the individual is relying on external objects and how they are manipulated and experienced (i.e., "seen" in a learned way). Spatial reasoning is also practiced by the very young (without language) and by people without sight.[5] All of these senses support spatial reasoning – and because we have different experiences, our spatial reasoning networks are unique and individual, as well as embedded in our shared cultures and shared experiences with the common objects and representations of our lives. Spatial reasoning becomes embedded in cultural and schooled professional practices (see Chapter 8), and like other rich reasoning can involve sequential chains and networks of cognitive activities. These activities involve long-term and working memory, executive functions and moves among representations and their transformations.

One challenge with spatial reasoning, as a form of school-based knowing, is that it is not easily reducible to alphanumeric modes of communication, and thus, it is not thought of as something that can be transmitted, like much of school mathematics is thought to be. In contrast, as a practice, a way of thinking in and of the world, it does not "get applied" to particular "real-world" situations, but rather reproduced in them. It often seems like key parts of spatial reasoning are internal to the individual, and can only be teased out through observation protocols or clever test items. However, there are cultural practices that can be learned within both the mathematical community and mathematics classrooms, which include working with models, diagramming and gesturing – all of which are based on and give rise to spatial reasoning.

Spatial reasoning in the work of mathematicians and within other spatialized domains

While spatial reasoning is not generally highlighted beyond selected areas of the mathematical sciences, it is frequently identified as a core element in how mathematics works (e.g., Thurston, 1995; Burton, 2004; Whiteley, 2005; Sinclair & Gol Tabaghi, 2010). Across such references, however, there is a tendency to talk about spatio-visual reasoning – that is, to conflate spatial reasoning and visual reasoning. The tendency is particularly prominent in areas of the mathematical sciences associated with geometry.

While geometry has been marginalized in most North American K–16 mathematics curricula, 3D geometry and the associated spatial reasoning is widespread over a number of applied areas (Whiteley, 1999; Clements & Sarama, 2011). As illustrated in Chapter 8, there is associated spatial reasoning, with varied disciplinary conventions, for representing and reasoning with spatial information. (A broader sweep of these critical areas of modern research in geometry can be found in Toth, O'Rourke, & Goodman, 2004.)

The experience of Whiteley (one of the authors) is cogent here, spanning several topics that involve motions and spatial planning in 3D space. Areas include: (i) computational geometry and robotics (in Computer Science); (ii) design and analysis of linkages (in Mechanical Engineering); (iii) control of formations of moving robots and localization of sensor networks (in Electrical Engineering and Computer Engineering); and (iv) modeling protein structures and their motions, with its impact on protein function and drug design. Such work has included developing physical models, running simulations over time, and preparing animations of how things unfold in time and space – all in support of spatial reasoning and communication within interdisciplinary collaborations.

More historically, there are rich areas of geometry involving spatial reasoning that have developed over millennia. Instances include the construction of circles and squares in ancient Vedic, Babylonian, and Greek altar construction, as well as the fitting together of 2D tiles in the stunning Islamic tessellations of the Alhambra. Work from the earliest "applied mathematician," Archimedes, includes extensive evidence of spatial reasoning. Examples include his exploration of the stomachion puzzle,[6] his derivation of the area of a parabolic segment, and his derivation of the volume of a hemisphere using what is now called Cavalieri's principle. This principle can be illustrated with a stack of coins (see Figure 1.2a), the volume of which must be constant whether organized in a right cylinder or a sheared stack. Cavalieri's principle is used, for example, to compare the area of a rectangle with that of a parallelogram having the same base and height (the rectangle transformed by shearing; see Figure 1.2b). This principle thus becomes a way of asserting whether or not two volumes or areas have the same measure, without having recourse to number or to algebra.

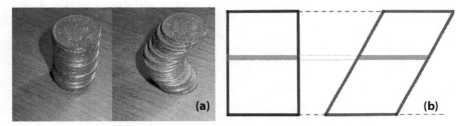

Figure 1.2: Illustrations of Cavalieri's principle.[7]

Linking ideas

We close this chapter with a brief recap of the contents and intentions of this book, beginning with overviews of how spatial reasoning has been characterized and studied by psychologists (Chapter 2) and mathematics educators (Chapter 3). Those chapters begin to frame the possibility of a dramatic rethinking of school mathematics.

With that view in mind, in Chapters 4 and 5 we begin to interrogate the structures and beliefs that might be holding current school mathematics in place. Chapter 4 offers an account of how current enactments of school mathematics emerged – focusing in particular on how they came to be so devoid of spatial reasoning. Chapter 5 extends this discussion, looking to "embodied" theories of knowing and learning. Those theories are used both to illuminate some troublesome-but-pervasive beliefs about school mathematics and to open up alternative ways to think about the project.

In Chapters 6, 7, and 8, we meld established research evidence and emergent pedagogical insights as we offer several classroom-based examples of what spatial reasoning might look like from pre-Kindergarten to Grade 3. In Chapter 8 we also offer examples from later studies which build on abilities developed on the early spatial reasoning. In our final chapter, we revisit and elaborate our construct of spatial reasoning, extend our discussion of a radically spatialized school mathematics, and offer some thoughts on possible foci for future research.

Spatial reasoning matters: to later careers, to support key learning, to feeling at home in our world, as a life skill. It matters across many fields including mathematics, and across many areas of mathematics and learning mathematics.

Notes

1   Spatial reasoning is also used as a selective filter in many domains, as evidenced through extensive uses of examinations such as the Medical College Admission Test (for North American Medical Programs, https://www.aamc.org/students/applying/mcat/), the Dental Admission Test (DAT: http://www.ada.org/en/education-careers/dental-admission-test/), the Mechanical Aptitude Tests, and associated Spatial Visualization Tests (http://www.aptitude-test.com/mechanical-aptitude.html).

Engineering programs in particular have identified spatial reasoning as a critical skill, especially for success in the first year and retention to the second (Sorby et al., 2013). While outside our specific interest in the early years, it is notable that undergraduate students who arrive with measurable deficits in spatial reasoning can effectively learn the skills necessary to succeed. Analogous gaps appear to be present at every level of formal education, and so such evidence signals a need for support materials, and qualified teachers to support the development of spatial reasoning.

2   Against the charge emerging from the Bourbaki-era of anti-imagery, Jean Dieudonné (1981) argued that "if anybody speaks of 'the death of Geometry' he merely testifies to the fact that he is utterly unaware of 90% of what mathematicians are doing today" (p. 231). More recently, Sir Michael Atiyah argued that "spatial intuition or spatial perception is an enormously powerful tool and that is why geometry is actually such a powerful part of mathematics – not only for things that are obviously geometrical, but even for things that are not. We try to put them into geometrical form because that enables us to use our intuition" (Atiyah, 2002, p. 6).

3   While we do not wish to insist on a strict duality between the elements in these pairings, they serve to illuminate the extent to which the first item in each dyad is preferred in current society.

4   For example, the psycholinguist Donaldson (1986) characterized explanation as "essentially a verbal activity" (p. 2), before carrying out systematic work with children (aged 3 to 11) on the central English causal connectives "because" and "so," which both mark and signal attributions of cause and effect (with opposite identifications of causality: X, *because* Y and X, *so* Y). However, the diagrammatic reasoning community has already contradicted this narrowing of the term with their analysis of alternative valid forms of human reasoning. The models of working memory have the phonological loop (language), the Visuo-Spatial Scratchpad and the Episodic Buffer, which can draw on both. See also Stenning (2002).

5   See Healy and Fernandes (2011) for an example of ways in which learners who are blind see with their hands.

6   The stomachion puzzle involves rearranging fourteen polygonal puzzle pieces into a square in order to make recognizable shapes, such as an elephant.

7   Figure 1.2a is drawn from the Cheswick Chap Creative Commons.

2

The development of spatial reasoning in young children

YUKARI OKAMOTO, DONNA KOTSOPOULOS, LYNN McGARVEY, DAVID HALLOWELL

— In brief …

Drawing primarily from the psychological literature, this chapter describes what we currently know about young children's development of spatial reasoning. We provide an overview of spatial reasoning as a multifaceted construct through concrete examples. This chapter provides a context for the chapters to follow.

Dimensions of spatial skills

Between birth and the time young children start schooling, rather remarkable changes in visual and spatial development have already occurred. Growth and development in children's spatial competencies are intertwined with their increasing capacity to move, navigate through, and interact with their environment. Although we know that children's spatial abilities are attributable to a complex and dynamic interaction of biological, psychological, sociological, and cultural influences, the vast majority of literature in this area is from a psychological point of view. In this chapter, then, we provide an overview of the ways in which young children's spatial reasoning has been studied as a multifaceted construct within psychological research. Viewing young children's spatial reasoning solely from the domain of psychology is only one part of the story, but we offer it here as a starting place for discussion. Familiarity with the spatial potentialities of young children provides us with a basis for discussing how spatial reasoning can contribute to children's mathematical understanding discussed in Chapter 3 and the value of spatial reasoning to children's futures in many STEM fields (Newcombe & Stieff, 2012; Uttal et al., 2013; see also Chapter 8).

To organize our discussion, we use the typological framework of spatial skills proposed by Uttal et al. (2013). The 2 × 2 classification scheme is based on two key dimensions of spatial reasoning: static versus dynamic skills and intrinsic versus extrinsic skills. In brief, the static–dynamic dimension distinguishes two types of spatial information – one that is fixed and the other that changes. We use static spatial skills when the main object of analysis or one's frame of reference remains stationary throughout the task; whereas dynamic skills are called upon when objects move or are moved. The intrinsic-extrinsic dimension distinguishes the nature of spatial information to attend to. We use "intrinsic skills" to define or describe an object whereas we use "extrinsic skills" to code objects' locations relative to each other or to a frame of reference. Figure 2.1 provides an overview of the two dimensions and the four resulting cells, and also shows an exemplar test that is associated with each cell in this classification scheme. Although this 2 × 2 scheme is not without issue, it provides an organizational structure by which to categorize and discuss the multitude of seemingly disparate skills and psychometric measures under the broad umbrella of spatial reasoning. In the following sections, we discuss what we know about children's spatial skills using this classification scheme. We first discuss intrinsic and extrinsic spatial skills that are considered static, then intrinsic and extrinsic skills that are dynamic.

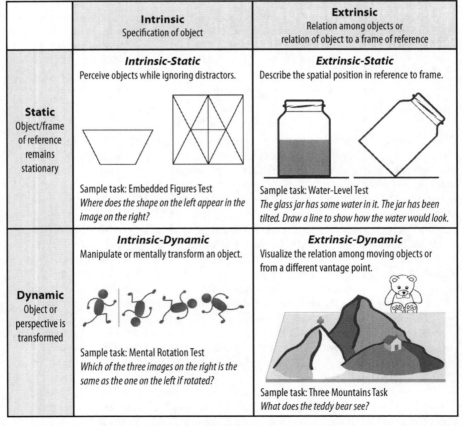

	Intrinsic Specification of object	Extrinsic Relation among objects or relation of object to a frame of reference
Static Object/frame of reference remains stationary	**Intrinsic-Static** Perceive objects while ignoring distractors. Sample task: Embedded Figures Test *Where does the shape on the left appear in the image on the right?*	**Extrinsic-Static** Describe the spatial position in reference to frame. Sample task: Water-Level Test *The glass jar has some water in it. The jar has been tilted. Draw a line to show how the water would look.*
Dynamic Object or perspective is transformed	**Intrinsic-Dynamic** Manipulate or mentally transform an object. Sample task: Mental Rotation Test *Which of the three images on the right is the same as the one on the left if rotated?*	**Extrinsic-Dynamic** Visualize the relation among moving objects or from a different vantage point. Sample task: Three Mountains Task *What does the teddy bear see?*

Figure 2.1: A 2 × 2 typology of spatial reasoning categories (adapted from Uttal et al., 2013).

Static spatial skills

We begin by examining current findings related to children's static spatial skills. In tasks requiring static spatial skills, the main object of analysis or one's frame of reference remains stationary. However, the tasks involve interpreting or perceiving spatial information based on potentially confusing or distracting information or images. The Embedded Figures Test (EFT) and the Water-Level Test (WL) are classic examples of how children's static spatial skills are assessed (see Figures 2.2 and 2.3). The psychological literature offers many other examples to highlight the typical developmental progression and potential conceptual or perceptual errors that young children make. We divide the discussion of static spatial skills into Uttal et al.'s (2013) intrinsic-static and extrinsic-static skill categories.

Intrinsic-static skills

Imagine examining the architectural drawings for renovations in your kitchen. Can you picture what your kitchen will look like once it is complete? If so, you are using spatial visualization skills classified in the intrinsic-static category. Visual artists, such as painters, photographers, sculptors, and interior designers tend to excel in intrinsic-static skills (Kozhevnikov, Hegarty, & Mayer, 2002; Kozhevnikov, Kosslyn, & Shephard, 2005). These skills are also associated with the classification and property analysis schemes of scientists (Hegarty et al., 2010; Mix & Cheng, 2012; also see Chapter 8).

Children and adults who show strong spatial skills in this category can perceive an object's spatial configurations against distracting background information. In the learning style literature, these individuals would be considered to have a field-independent learning style – a tendency to separate details from the surrounding context. Intrinsic-static skills are necessary for recognizing, describing, and classifying the spatial attributes of an object and the relation of parts within the object (Kastens & Ishikawa, 2006). Figure 2.2 (on the next page) provides examples of common tests used to assess intrinsic-static skills.

The two most common measures in the category of intrinsic-static skills are the Embedded Figures Test (EFT; Witkin, 1971) and the Hidden Figures Test (Ekstrom et al., 1976). Witkin's EFT, for example, asks participants to identify simple shapes that are embedded in a complex shape. The original test has been modified for use with children 5 to 12 years old (Children's Embedded Figures Test; Karp & Konstadt, 1971) as well as with preschoolers 3 to 5 years old (Preschooler Embedded Figures Test; Coates, 1972). The latter measure has often been used to test children's field-independence (e.g., Busch et al., 1993).

Studies focusing on preschool children's intrinsic-static spatial understanding are scarce. As Newcombe and Shipley (in press) noted, this may be due to a lack of reliable measures for children of this age group. Researchers are beginning to develop such measures. For example, at the Spatial Intelligence and Learning Center (SILC) – a National Science Foundation *Science of Learning Center*, several measures including those for the intrinsic-static category are under construction. These include measures to assess children's

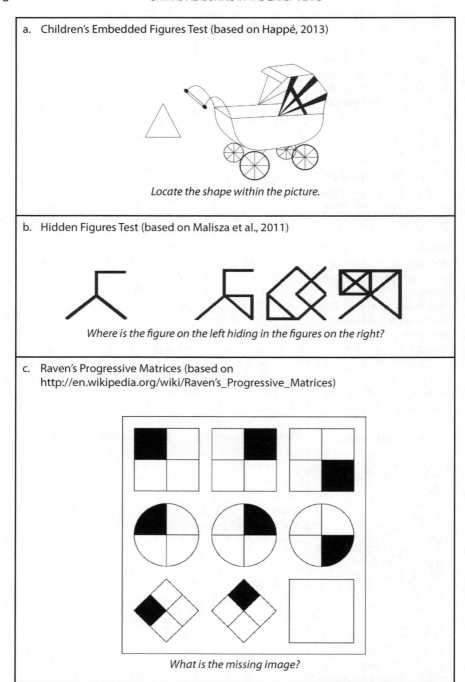

a. Children's Embedded Figures Test (based on Happé, 2013)

Locate the shape within the picture.

b. Hidden Figures Test (based on Malisza et al., 2011)

Where is the figure on the left hiding in the figures on the right?

c. Raven's Progressive Matrices (based on http://en.wikipedia.org/wiki/Raven's_Progressive_Matrices)

What is the missing image?

Figure 2.2: Tests of intrinsic-static skills.

recognition of basic 2D and 3D geometric shapes as well as their comprehension of spatial terms (Newcombe & Shipley, in press).

Although valid and reliable measures for preschool children have yet to be developed, some findings are emerging that document how young children perceive and manipulate intrinsic-static information. For example, Clements et al. (1999) examined young children's recognition of basic 2D geometric shapes. They were interested in finding out what criteria children use to decide if a particular shape is a circle, square, triangle, or rectangle. For a circle task, for example, children saw several different size circles drawn on a paper, along with distractors whose features included some roundness but were not circles. They were then asked to find all circles on the page and explain why they thought a particular shape was or was not a circle. This was repeated for the three other geometric shapes as well as one complex configuration of overlapping shapes. Children's explanations included visual type (e.g., "looks like a circle") and property type (e.g., "no corners"). In terms of accuracy, preschoolers were able to identify circles more accurately but their accuracy declined for squares, triangles, and rectangles in this order. Both 4- and 5-year-olds were correct only about half the time when asked to identify rectangles. Analyses of children's explanations indicated that preschoolers tended to use visual, not property cues to identify geometric shapes. These results suggest that preschoolers can identify familiar shapes (with varying degrees of accuracy) but that their initial criterion for identification is to match based on visually salient features of shapes.

The skills we examined in this section pertained only to individual objects whose spatial location or reference points do not undergo change. We now examine extrinsic-static skills that involve understanding inter-object representations and relations.

Extrinsic-static skills

Imagine standing in front of a large subway route map in an unfamiliar city as you try to get from point A to point B. Which route or combination of routes amongst a myriad of possibilities might you take to your destination? Extrinsic-static skills, such as planning a route, predominantly consider mapping or other perceptual tasks. Figure 2.3 (on the next page) shows sample tasks that have been used in the literature to measure extrinsic-static skills. In this set of tasks, children are to code the object's location in relation to a reference system against a distracting field change. According to Uttal et al.'s (2013) meta-analysis, extrinsic-static skills have the greatest potential to be learned through education and training (effect size of 0.69) compared to the three other categories of skills.

The classic examples of extrinsic-static perceptual tasks used with children include spatial tests of horizontal (see Water-Level Test in Figures 2.1 and 2.3a) and vertical invariance (see Plumb-Line Test in Figure 2.3b and Rod and Frame Test in Figure 2.3c). These tasks are based on Piaget and Inhelder's (1967) classic, but now extensively critiqued (see Sarama & Clements, 2009 for a general discussion) theory of young children's spatial perceptions. Piaget's work, along with the multitude of studies that followed, offered evidence of young children's perceptual abilities and difficulties with 2D and 3D

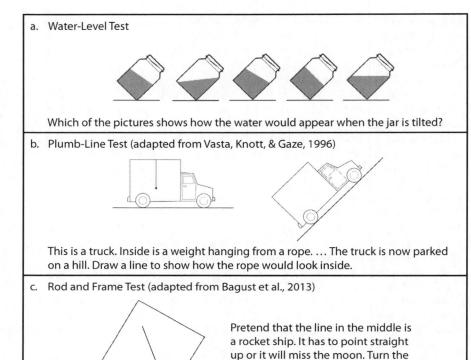

a. Water-Level Test

Which of the pictures shows how the water would appear when the jar is tilted?

b. Plumb-Line Test (adapted from Vasta, Knott, & Gaze, 1996)

This is a truck. Inside is a weight hanging from a rope. ... The truck is now parked on a hill. Draw a line to show how the rope would look inside.

c. Rod and Frame Test (adapted from Bagust et al., 2013)

Pretend that the line in the middle is a rocket ship. It has to point straight up or it will miss the moon. Turn the rocket ship so it will go straight up.

Figure 2.3: Tests of extrinsic-static skills.

horizontality and verticality tests. More recent evidence suggests that children's extrinsic-static skills may be better when using real or 3D models than 2D representations of them (Baldy, Devichi, & Chatillon, 2004).

A related task of verticality is the Rod and Frame Test (Witkin & Asch, 1948). The test has been used with children as young as 3 years of age. The test examines whether children can vertically align a line within the outline of a skewed square that serves as a distracting visual frame of reference. Early investigations noted that children tended to produce alignment errors, which decreased with age (Bagust et al., 2013).

Shifts in successful performance on verticality and horizontality tasks coincide with when children are first able to make decisions about field features and distractors (Pascual-Leone & Morra, 1991). Young children are highly influenced by references internal to the object rather than external cues. For example, in the study to examine children's verticality, Baldy et al. (2004) used a 3D model of a hill with a house sitting on a hilltop and asked children to place electric poles on both sides of the house to light a house. Baldy et al. found that children performed better when blindfolded than in the visual condition. That is, being unable to see the task setting minimized the field distractors and enhanced performance, suggesting that when field distractors are minimized, performance increases. Children's ability to consistently apply rules and strategies (e.g., a pole cannot lean over on the hill or it will fall over) increases and assists them with mitigating field distractors, which continue to be highly influential (Baldy et al., 2004).

The development of extrinsic-static information in young children has also been studied with simple maps and models. Uttal (2000) suggested that a young child's ability to understand maps and surveys initially emerges from an egocentric perspective – things known and important to the child, and also relative to the present visual field of the child. Teaching can influence this set of skills. Three-year-olds are able to use geometric cues that are explicit (e.g., defined lines rather than inferred shapes) for locating objects on maps (Vasilyeva & Bowers, 2006). They also show an early sign of the ability to identify the location of an object in a 2D representation and its corresponding location in an enclosed room or small space using relatively simple maps (DeLoache, 1987; Huttenlocher, Newcombe, & Vasilyeva, 1999). Within another couple of years, children begin to identify object location through inference such as using combinations of objects to form geometric cues or partially constructed geometric objects (Vasilyeva & Bowers, 2006).

Some researchers have pointed out that children's interpretation of maps is related to an ability to use a combination of features and landmarks (Cheng, Huttenlocher, & Newcombe, 2013). Others have emphasized that it is related to the ability to use geometric properties such as length or distance and angle (Lee, Sovrano, & Spelke, 2012). What is undisputed is that children become more proficient in understanding maps and models with age and this may be attributed to a variety of factors including more experience with maps (Uttal, 2000), broader environmental experiences (Piaget, Inhelder, & Szeminska, 1960; Piaget & Inhelder, 1967), and conceptual changes related to abstraction of space (Clements, 2004).

Dynamic spatial skills

We next examine current findings related to children's dynamic spatial skills. Dynamic spatial skills are most closely studied and frequently associated with STEM fields. In contrast to understanding characteristics of static objects, dynamic spatial skills often involve transforming an object or set of objects, perhaps by rotating, folding, bending, or scaling. A related set of skills refers to perspective taking tasks, which require a real or imagined shift of one's personal frame of reference. The classic psychometric measures in this area include the Mental Rotation Test (Figures 2.1 and 2.4a–c) and variations of Piaget's Three Mountains Task (Figures 2.1 and 2.5a). Many additional examples are provided below as we distinguish intrinsic-dynamic and extrinsic-dynamic skills in the next two sections.

Intrinsic-dynamic skills

You notice that your child spends hours playing Tetris or similar video games requiring mental rotation. Or imagine your students are trying to copy a teacher's design of a Lego™ car with a motor and simple drive system. The students are analyzing how individual pieces from the motor to the gears to the axle to the wheels need to be oriented to fit together properly, visualizing how the pieces will operate under rotation. Activities such as these require intrinsic-dynamic skills, which one day might transfer over to a career as a scientist, mathematician, or engineer. These intrinsic-dynamic skills concern some spatial transformation of a given object such as rotation, folding, cross-sectioning, and deformations (Newcombe & Shipley, in press). In addition to the measures mentioned above, this type of spatial reasoning is associated with a number of measures, including block designs, paper folding, and the Wheatley Spatial Test (Wheatley, 1996; Uttal et al., 2013). Figure 2.4 provides different types of tasks in this category.

Intrinsic-dynamic spatial reasoning is another domain that warrants attention from educators, as training has been found to improve intrinsic-dynamic spatial ability (see Chapter 3). Uttal et al.'s (2013) meta-analysis found that training interventions regardless of age resulted in significant improvement to intrinsic-dynamic spatial reasoning (effect size 0.44). There was some suggestion that interventions were more successful with young children and that in all age categories training was equally effective for females and males.

The literature on intrinsic-dynamic mental transformations has documented a U-shaped pattern of development in which intrinsic-static abilities emerge in infancy (as early as 3 months old), only to disappear temporarily at around age 3 or 4, and re-emerge after 5 years of age (Frick & Wang, 2014). Recent findings suggest, however, that preschoolers do succeed when given age-appropriate measures of mental rotation (e.g., Krüger et al., 2014; Hawes et al., in press). At present, researchers continue to develop spatial measures appropriate for preschool-age children (e.g., Spatial Intelligence and Learning Center Tests and Instruments, 2014).

Recent research suggests that gender differences in spatial reasoning have been overstated. However, mental rotation appears to be one area of intrinsic-dynamic spatial reasoning that has repeatedly demonstrated a male advantage (Newcombe & Stieff, 2012; Frick, Möhring, & Newcombe, in press). These tasks require one to mentally rotate plane objects such as polyomino shapes or solid objects such as multilink cubes in order to select the rotated shape from an array of target items. Timed conditions on this task have consistently shown gender differences. However, performance differences do not completely disappear in untimed conditions (Voyer, 2011).

Some research suggests that this performance difference may be related to the fact that females prefer a careful stepwise process to analyzing shape orientation, whereas males tend to rely on automatic processes that are closer to the kind of mental operations used in everyday perception (Jaušovec & Jaušovec, 2012). Teachers might help close the gender gap in performance by ensuring that female students are getting ample experience with these kinds of tasks in low-anxiety contexts. While future research is required to answer the question definitively, it is not unreasonable to expect that as female students gain familiarity, experience, and confidence with mental rotation tasks, their cognitive strategies might become automatic and diminish gender differences in performance in this area. Also, it is important to remember that this

a. Analogue Mental Rotation Task (based on Krüger et al., 2014)

Straighten out the teddy bear so that the two pictures look the same.

b. 3D Mental Rotation Block Task (based on Hawes et al., 2015)

Which shape looks like the target item?

c. Two-Dimensional Mental Transformations: Children's Mental Transformations Task (based on Ehrlich, Levine, & Goldin-Meadow, 2006)

Choose which arrangement on the right could be made by bringing the images on the left together?

d. Two-Dimensional Mental Paper Folding: Mental Folding Test for Children (based on Harris, Hirsh-Pasek, & Newcombe, 2013)

If you fold the paper on the left as shown, what would the folded sheet look like?

e. Mental Cutting Test (based on http://www.silccenter.org/index.php/resources/testsainstruments)

Which of the four shapes on the right could be made by cutting through the object on the left?

Figure 2.4: Tests of intrinsic-dynamic skills.

difference is in regard to average group performance. On the individual level of comparison, many female students outperform male students on these tasks. Mere awareness of stereotypes alone has been shown to significantly hinder student performance when students are concerned they may validate the stereotype (Spencer, Steele, & Quinn, 1999). Thus it is important not to create spatial anxiety in the classroom by focusing on group differences.

An area of mathematical relevance is the work of Clements, Wilson, and Sarama (2004) on shape composition and decomposition. Tasks used in this work involve using pattern blocks to compose a person-like figure given an outline. Success at this task requires children to predict a resulting shape when geometric shapes are put together to form composite shapes. As children matured, they were seen to progress along a trajectory from initially being unable to combine geometric shapes in order (precomposer), to succeeding using trial and error (piece-assembler), to beginning to attend to features of shapes to make the image (picture-maker), and finally to using mental imagery and shape properties to complete shape compositions (shape-composer). These skills are thought to underlie future success in geometry, and other works have shown that early training in the spatial domain extends to improved symbolic understanding of number developmentally (Gunderson et al., 2012). Teachers in early grade levels might consider regular time with tangram puzzles, providing additional assistance to students who struggle but do not seem to notice important features for filling puzzle frames.

Preschoolers' understanding of 3D shapes has been studied less extensively, but we can draw insights from an early study by Reifel and Greenfield (1983) on block building. Reifel and Greenfield read a story to children and asked them to use blocks to create any objects that appeared in the story. They found that young children initially had difficulty and used a limited number of blocks to create a new object in the story (e.g., house). This finding appears to be in agreement with the studies of 2D pattern blocks in that 3D block building also requires spatial visualization skills (Linn & Petersen, 1985).

Extrinsic-dynamic skills

Imagine a solar system model where the planets rotate around the sun, moons rotate around the planets, each of these bodies rotates on an axis, and your job is to predict when a solar, lunar, or planetary eclipse will occur and from where it might be visible. Extrinsic-dynamic skills are at play in this example that involves multiple bodies moving in and through space, where spatial relations among objects are constantly in flux, and the perspective changes from one location to another. The extrinsic-dynamic category involves recognizing the dynamic or changing spatial relations among two or more objects or between one's own moving body and objects or landmarks in the environment. There are two types of navigational skills within this category: self-to-object and object-to-object navigation. See tests of extrinsic-dynamic skills in Figure 2.5.

The first, self-to-object navigation, involves reasoning about one's own movement through space, particularly by referencing one's body to objects located in one's changing environment. For example, a child who knows there are multiple ways to get to the park depending on whether they are at home,

a. Perspective Taking Test for Children (based on Frick, Möhring, & Newcombe, 2014)

Lisa (the doll in the first picture) took a photo. Point to the picture that Lisa took from where she was standing.

b. Place Learning: Object-to-object orientation

See, I put a silver coin in the sandbox. I'm going to cover it up, spin you around, and take you to the other side. Where is the coin hidden?

Figure 2.5: Tests of extrinsic-dynamic skills.

at school, or at a friend's house is using self-to-object navigation skills. These skills include self-locomotion and also perspective taking (Figure 2.5a). Motor development is particularly influential in a child's spatial cognition from infancy. Each developmental milestone in the domain of motor skills – from sitting, to crawling, to walking – influences children's subsequent actions upon and with their environment. Interestingly, in tasks involving infants' searching capabilities, experienced crawlers and walkers are more successful at finding toys in large-scale environments than are novice crawlers and walkers (Clearfield, 2004; Berger, 2010). Locomotive experience, rather than age, appears to be of tremendous importance. In general, as soon as children begin to move on their own, they begin to demonstrate path integration; that is, they start to encode direction and distance with components of bodily translation and rotation in order to navigate and update their bodily position in space.

The research on self-to-object navigation focuses not only on the actual physical movement through space, but also the imagined movement from one location to another, requiring the navigator to visualize the spatial environment from a different perspective. A classic assessment task in this area is Piaget's Three Mountains Task, along with its multitude of variations. In this task, a child is seated in front of a 3D model of three mountains, each mountain featuring different attributes (Figure 2.1). A doll is seated at the opposite end. The child is asked to describe the view from the doll's vantage point. Although the task has been critiqued in subsequent studies for the way it was administered (e.g., Rieser, Garing, & Young, 1994), it inspired a large body of research on perspective taking. Children as young as 18 months demonstrate an awareness that a clear line of sight is necessary for another person to see what the child sees, and are also beginning to understand that another person may see something different from the child when viewing a scene from a different perspective (Flavell, 1999). Under some conditions, preschool children recognize that another person may see the same set of objects, but may see them differently if viewed from a different location (Moll et al., 2013). Conflicting findings, however, have been reported in a recent study that shows that a full capacity to take spatial perspectives of others does not emerge until about age 8 (Frick et al., 2014; see also Figure 2.5a). There is much more to be studied about children's perspective taking skills and the relation of this skill to mathematical thinking. Perspective taking activities provide children with opportunities to imagine and describe models or environments from various perspectives, or from changed positions such as a playground or a hallway of classrooms (e.g., see the navigation activity outlined in Chapter 7). Children's early understanding of extrinsic-dynamic operations begins to develop very early in life, grows throughout the early years, and varies among individuals based on experiences.

The second form of navigation, object-to-object, involves updating spatial relations when one or more objects are in motion. Although this aspect of extrinsic-dynamic skills does not have an extensive body of research, even young children are said to have geometric knowledge of object-to-object spatial relations (Trafton & Harrison, 2011). An area that has been studied with young children is described as place learning. In these tasks, children use direction and distance of landmarks and spatial-geometric properties to search for and locate hidden objects. A common task to assess place learning involves not only hiding objects, but also taking children to the opposite side of the room or around the object that has been hidden to locate the object (see Figure 2.5b). From this new vantage point, children need to take into consideration the shifts in spatial relations between the hidden object and the landmarks that may be used to locate the object (Newcombe et al., 1998). Mathematically, the potential skills involved in place learning include direction, distance, and angle in 2D surfaces and 3D environments.

In both systems of navigation in dynamic contexts, spatial language plays a key role. Recent research suggests that children's spatial language use is related to their parents' language and that children's spatial language is a predictor of future spatial skills (Pruden, Levine, & Huttenlocher, 2011). The early development of spatial language involves prepositions and verbs – such as *through, around, across, over,* and *toward* – may be used to describe the child's body through space or dynamic object-to-object spatial relations (also see Chapter 7). The use of spatial language and gesture, and the role they play in spatial cognition, is a promising new field of study with much to be learned.

The same might be said of explicitly and intentionally increasing spatial vo-cabulary among students, and encouraging students to gesture while explain-ing their spatial reasoning during classroom activities.

Linking ideas

In this chapter we explored spatial reasoning through a psychological lens where an extensive body of literature has been developed over the course of more than half a century. As was noted in Chapter 1, there has been con-siderable debate over the multitude of constructs and terms associated with spatial reasoning, particularly in the field of psychology. Our use of Uttal et al.'s (2013) typological framework of spatial skills allowed us to loosely cat-egorize and outline key findings and relevant measures of young children's spatial reasoning along its static versus dynamic and intrinsic versus extrinsic dimensions.

As we examined the literature relevant to young children's spatial rea-soning we recognized that the skills and tasks did not always fit neatly into one category or another. Indeed, skills shifted from one category to another depending on the interpretation of the task at hand or ways in which a single task is performed. For example, a geometric model may be considered as one object (intrinsic) with multiple internal parts or as comprising of multiple ob-jects (extrinsic). A task that involves rotation of the model may be classified as intrinsic-dynamic if it is considered to involve mental rotation. If children are to imagine viewing the model from a different vantage point, the skill is con-sidered to be extrinsic-dynamic. These slight variations become problematic because the task design predetermines the category rather than the skill. If children were asked more generally to describe what the other side of a geo-metric model might look like, some children may mentally rotate the model while others might imagine moving their bodies to view the model from the other side. In this case, two responses would be classified differently.

Although attempting to categorize the sets of discrete skills and tasks proved challenging in this chapter, the limitations of the framework become even more apparent in Chapter 3. Rich tasks given in classroom contexts often span multiple categories (Bruce & Hawes, in press). What we know about early spatial reasoning has emerged predominantly from psychological stud-ies where the research findings are almost exclusively a result of laboratory or experimental settings using props and methodologies that embody Western cultural biases. As we point to in the final chapter of this book, research is needed that examines spatial reasoning from social and cultural domains, as well through other areas of cognitive science including perception (e.g., Seung & Lee, 2000) and embodiment (see Chapter 5). Additionally, the 2 × 2 typol-ogy tells us very little as to what spatial activities are relevant to mathematics, which would support the work of teachers in early childhood classrooms.

Despite its limitations, the 2 × 2 classification matrix allowed us to con-sider the necessary depth and diversity of spatial reasoning that is often ne-glected in the mathematics curriculum. In particular, early-years mathematics tends to be substantially limited to static skills such as sorting and noticing properties of shapes and patterns. Yet, the research we reviewed here demonstrates young children's remarkable capabilities in other areas. These points are further taken up in subsequent chapters (Chapters 3, 6, and 7). Furthermore, our discussion also points to the necessity of attending to spatial reasoning in the early years given the strong connection between spatial skills, such as mental rotation, and its paramount importance to STEM fields (Chapter 8).

3

Developing spatial thinking

ZACHARY HAWES, DIANE TEPYLO, JOAN MOSS

In brief ...

Once believed to be a fixed trait, there is now widespread evidence that spatial reasoning is malleable and can be improved in people of all ages. In this chapter, we first discuss the relationship between spatial reasoning and mathematics and then present a variety of spatial training approaches shown effective in not only supporting children's spatial reasoning but mathematics performance as well.

Spatial reasoning as a foundation for mathematics learning

There is an emerging consensus that spatial reasoning plays a foundational role in the early development of mathematics. Due in part to the recent design of age-appropriate measures of spatial reasoning for young children (see Chapter 2, Figure 2.1), researchers have begun to understand how early spatial skills relate and contribute to the learning of school mathematics. In a longitudinal study that followed children from the ages of 3 to 5, Farmer et al. (2013) found evidence to suggest that children's spatial skills at 3 years of age were strong predictors of how well the same children performed in mathematics two years later, upon formal school entry. Moreover, spatial skills were better predictors of later mathematics performance than vocabulary and even mathematics.

In another study, Verdine et al. ("Finding the Missing Piece," 2014) reached a similar conclusion. Spatial skills assessed at the age of 3, along with executive function skills assessed at the age of 4, predicted over 70% of the variability in mathematics performance at 4 years of age. Even after controlling for the contribution of executive functions, spatial skills predicted 27% of the variability in children's mathematics performance. It is worth noting that in both of the above studies, the researchers used a relatively simple means to assess children's spatial reasoning. Children were presented with Mega Bloks™ arrangements and asked to copy them as accurately as possible (see Figure 3.1). A score was assigned to each child based on how accurately he or she was able to replicate the design.

Figure 3.1: Stimuli for the Test of Spatial Assembly
(TOSA; see Verdine et al., "Finding the Missing Piece," 2014).

We mention how the researchers assessed spatial skills because we think it is relevant for the consideration of early classroom interventions. Indeed, there is sufficient evidence suggesting that early construction skills – of the sort that involves copying, drawing, and block building – play an important role in the learning of mathematics (Casey, Andrews, et al., 2008; Casey, Erkut et al., 2008; Tzuriel & Egozi, 2010). For example, Wolfgang, Stannard, and Jones (2001) carried out a longitudinal study that followed children from preschool to adulthood (a period spanning 16 years). The researchers showed the complexity of block building at age 5 was a significant predictor of how well the same individuals performed in high school mathematics.

Taken together, the above research findings suggest that spatial reasoning and mathematics are co-related and that early spatial skills may provide a foundation on which mathematics learning is built. This raises the question of how and why mathematics and spatial reasoning are related.

How and why spatial reasoning helps "do" mathematics

The question of how and why spatial reasoning and mathematics are linked remains largely unknown. In their recent review on the subject, Mix and Cheng (2012) urged the field to move beyond correlational studies, stating:

> The relation between spatial ability and mathematics is so well established that it no longer makes sense to ask whether they are related. Rather, we need to know why the two are connected – the causal mechanisms and shared processes – for this relation to be fully leveraged by educators and clinicians. (p. 206)

The last part of their statement is a particularly important point. In order to fully harness and develop the powers of spatial reasoning in our mathematics classrooms, we need to have a strong theoretical stance and evidence-based knowledge as to why the two go hand in hand. As mathematicians, mathematics educators, teachers and curriculum developers, we need to work together to understand more about the connections between spatial reasoning and mathematics. Indeed, teachers need to be able to recognize and theorize when spatial skills are needed to support mathematics learning, as well as when a focus on number might hinder or prevent mathematical understanding (Newcombe, 2014).

Perhaps an example better serves this point. A teacher would have little difficulty explaining to a curious parent why so much time was being spent on developing number sense. The practicality is self-evident; numbers are ubiquitous and endlessly useful and for this reason, it is unlikely that a parent would even ask such a rhetorical question. However, if asked to explain why so much time was being spent (in math class nonetheless!) on developing spatial reasoning, a teacher would likely be facing a much more difficult challenge. The relationship between spatial reasoning and mathematics is not always immediately apparent, and yet, decades of research inform us that the two are intimately connected. Based on prior research, and our own experiences working in classrooms, we offer three reasons why spatial reasoning is related to and helps support the learning of mathematics.

Mathematics as an inherently spatial subject

There is an abundance of examples, in which to "do," "create," and "express" mathematics is to use and depend on spatial reasoning and spatial representations. As mentioned in Stanislas Dehaene's book, *The Number Sense* (2011), it is "almost as if they [spatial reasoning and mathematics] were one and the same ability" (p. 135). Clements and Sarama (2011) posit that it is through mathematics that we "communicate ideas that are essentially spatial. From number lines to arrays, even quantitative, numerical, and arithmetical ideas rest on a geometry base" (p. 134). Indeed, in our own work in early years classrooms, we are regularly confronted with examples of how spatial reasoning and mathematics are intimately linked. Linear and area measurement, early patterning and algebra, fractions, symmetry, and not to mention geometry, are inextricably linked to children's understandings of spatial relationships. Even something as simple as comparing shapes or numbers becomes an act of spatial reasoning when the objects assume different orientations. Interestingly, research suggests that the role of spatial reasoning and the use of spatial representations become even more important as one advances in their learning of mathematics (Mix & Cheng, 2012). In the following quote we are reminded to continually pay attention to the highly visual and spatial nature of calculus.

> The role of visual thinking is so fundamental to the understanding of calculus that it is difficult to imagine a successful calculus course which does not emphasize the visual elements of the subject. This is especially true if the course is intended to stress conceptual understanding, which is widely recognized to be lacking in many calculus courses as now taught. Symbol manipulation has been overemphasized and in the process the spirit of calculus has been lost. (Zimmermann, 1991, p. 136)

It is easy to lose sight of the importance of spatial reasoning in mathematics. The representations used in spatial reasoning are often private or internal to the individual learner, and as such, are often difficult to externalize and share through external community conventions, or rather lack thereof (Whiteley, 2014). In many ways, spatial reasoning is so much a part of mathematics that we take it for granted, we forget to acknowledge its role, and we do little to harness its potential (see Clements & Sarama, 2004).

Numbers are represented spatially

For over a century now, researchers have revealed a close relationship between space and numbers (Galton, 1880; Mix & Cheng, 2012). Dating back to the late 1800s, Sir Francis Galton provided anecdotal evidence that for some, individual numbers were seen in the "mind's eye" as objects that occupied distinct visual and spatial forms:

> Those who are able to visualize a numeral with a distinctness comparable to reality, and to behold it as if it were before their eyes, and not in some sort of dreamland, will define the direction in which it seems to lie, and the distance at which it appears to be. If they were looking at a ship on the horizon at the moment that the figure 6 happened to present itself to their minds, they could say whether the image lay to the left or right of the ship, and whether it was above or below the line of the horizon; they could always point to a definite spot in space, and say with more or less precision that that was the direction in which the image of the figure they were thinking of first appeared. (1881, p. 86)

These "number forms," as Galton referred to them, provided one of the earliest accounts of a suspected link between numerical and visual-spatial processes. Galton noted that the experience of number forms was a relatively stable trait within individuals, but large variation existed between individuals. The visual-spatial properties associated with number forms varied according to spatial orientation, color, brightness, and perceived weight (see Figure 3.2; Galton, 1880, 1881). Taken together, this work suggested that numbers were internally represented as objects and occupants of distinct positions in linear space (Galton, 1880; de Hevia, Vallar, & Girelli, 2008).

During the last several decades, there has been resurgence in the scientific study of how humans mentally represent numbers (cf., Dehaene, Bossini, & Giraux, 1993). There is now extensive support for Galton's intuitions about the visual-spatial nature of numerical representations (Seron et al., 1992; de Hevia et al., 2008). While only a small segment of the population (approximately 12%) experience the vivid number forms described by Galton, the vast majority

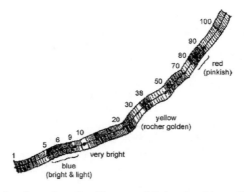

Figure 3.2: A number form described by one of Galton's subjects (from Galton, 1880).

unconsciously represents numbers spatially (Sagiv et al., 2006). For example, numerous studies show an automatic association of small numbers as belonging to the left side of space and larger numbers as belonging to the right side of space, a finding referred to as the SNARC effect (Spatial-Numerical Association of Response Codes; Dehaene et al., 1993). As such, people are faster to respond to smaller numbers (e.g., 1, 2, 3) with their left hand and faster to respond to numbers of larger magnitude with their right hand (e.g., 8, 9, 10). The reverse is true in societies that write and read numbers from a right-to-left orientation, such as the case in Palestine (see Shaki, Fischer, & Petrusic, 2009).

The influences of spatial representations of number are also present during simple arithmetic (Fischer & Shaki, 2014). In a recent special issue on the topic, a collection of findings demonstrated that not only are single digits subject to spatial biases, but arithmetic and even the operators themselves (i.e., plus (+) and subtraction (–) symbols) are associated with space (Fischer & Shaki, 2014). Addition problems elicit "right-side-of-space" biases whereas subtraction problems elicit "left-side-of-space" biases. Evidence of such spatial biases can be seen in the tendency for people to overestimate the result of addition problems and underestimate the result of subtraction problems, an effect referred to as the operational momentum (OM) effect (Fischer & Shaki, 2014). Werner and Raab (2014) discovered a link between the direction of people's eye movements and type of problem they were solving. The authors revealed a shift in attention toward the right side of space for addition solutions and a shift in attention toward left side of space for subtraction solutions. Together, these and other findings (see Fischer & Shaki, 2014) are thought to reflect the cognitive representation of magnitude meaning along a metaphorical "mental number line."

Other examples of how the "mental number line" might be implicated during mathematics come from studies from the psychology literature that utilize actual number lines. Over the past decade, there has been an explosion of research on the use and implications of findings related to number line estimation tasks. In a typical number line task, participants are presented with a line with only two end points (e.g. 0–10, 0–100, 0–1000; see Figure 3.3). Participants are then presented with a number and asked to indicate its exact location on the line. Performance on the task is thought to reflect the precision of an individual's mental number line or mental counting line. Importantly, performance on the task has been found to strongly predict concurrent and later mathematics performance (Siegler & Booth, 2004; Booth & Siegler, 2008). The findings of Siegler and others is that with age, experience, and training,

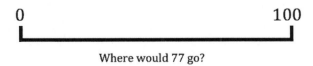

Where would 77 go?

Figure 3.3: Example item from a number line estimation task.

children's number line performance improves as a function of more accurate mappings of numbers to space.

Further evidence of a link between a spatial representation of number and arithmetic performance comes from Gunderson et al. (2012). In a longitudinal study involving two data sets, Gunderson et al. (2012) found that spatial skills (i.e., mental transformation skills involving rotation and translation) at the beginning of 1st and 2nd grades predicted growth in linear number knowledge (as assessed with a number line estimation task) over the course of the school year. Furthermore, children's spatial skills at age five predicted how well these same children performed three years later on an approximate symbolic calculation task. Interestingly, this relationship was mediated by children's number line performance during 1st grade. This finding is significant in that it suggests that spatial skills play an important role in the development of young children's spatial representations of number. Taken together, this line of research suggests that mathematics and spatial reasoning are tied together through the spatial representation of numbers. Preliminary research attempting to leverage this connection to improve number sense performance is encouraging. Research has shown that playing linear number board games for even one hour can increase at-risk preschool students' abilities to make number line estimations, and judge and compare numerical magnitudes (Siegler & Ramani, 2008, 2009).

The human tendency to represent numbers spatially is further supported by research in neuroscience. It is now widely recognized that both numerical and visual-spatial tasks require and depend on the recruitment of highly similar brain regions, namely, various neural networks located within the parietal cortex (de Hevia et al., 2008; Hubbard et al., 2009). For example, the spatial task that involves mental rotation and mathematical tasks requiring numerical processing, are both thought to rely on the intraparietal sulcus located within the parietal lobe (Hubbard et al., 2009). Indeed, Hubbard et al. (2009) suggest that "... the parietal mechanisms that are thought to support spatial transformation might be ideally suited to support arithmetic transformations [e.g., calculations] as well" (p. 238).

The above evidence suggests that spatial and numerical processes are closely linked and that space is a useful metaphor for how we think about numbers (Hubbard et al., 2009). However, more research is needed to reveal the specific mechanisms that underlie this relationship, and furthermore, to elucidate how spatial representations of number are related to mathematics more generally, including areas of mathematics that extend beyond simple arithmetic and calculations.

Mathematics and spatial reasoning involve visual-spatial working memory

Another way that spatial reasoning and mathematics might be linked is through shared cognitive resources, including the ability to mentally manipulate visual-spatial information. Visual-spatial working memory appears to be especially important for the early learning of mathematics. Children who have an easier time remembering and mentally manipulating visual-spatial information tend to have an easier time doing mathematics (Kyttälä et al., 2003; Kyttälä & Lehto, 2008; Thompson et al., 2013). Given this basic relationship, some researchers have found evidence (albeit somewhat controversially) that training-induced improvements in working memory (e.g., through computerized training exercises) result in improved mathematics performance (Holmes, Gathercole, & Dunning, 2009; St. Clair-Thompson et al., 2010; Witt, 2011). For the time being, these findings suggest a cause–effect relationship whereby improvements in working memory can be expected to aid in the performance of mathematics tasks.

Recall that it is intrinsic-dynamic spatial reasoning that has been, to date, most associated with performance in mathematics. Mix and Cheng (2012) hypothesized that the strength of this relationship depends on the shared demands placed on visual-spatial working memory. That is, both mathematics and intrinsic-dynamic spatial reasoning require the active maintenance and manipulation of visual-spatial information in one's mind. Therefore, it is possible that classroom interventions aimed at developing intrinsic-dynamic spatial reasoning might also strengthen children's visual-spatial working memory capacity – a core cognitive skill involved in the learning of mathematics. Future research efforts are needed to test this possibility along with a more detailed account of the role of visual-spatial working memory in spatial reasoning.

In returning to the question of why spatial reasoning matters for mathematics, we have presented evidence to suggest that (i) many mathematics tasks use inherently spatial representations (e.g., linear and area measurement, visualizing multiplication as arrays, geometrical transformations); (ii) numbers are represented spatially; and (iii) both mathematics and spatial reasoning involve visual-spatial working memory. In the next section, we examine the malleability of spatial reasoning and what this means for mathematics learning, and consider how these three accounts might be utilized in the design of effective classroom interventions that aim to bridge the mathematics and space divide.

Spatial reasoning can be improved through practice and targeted interventions

The strong link between spatial reasoning and mathematics raises the possibility that improving children's spatial skills might serve as a way to strengthen mathematics learning. To date, however, very few studies have pursued this line of inquiry. Uncertainty about the malleability of spatial reasoning may be one reason for this. After all, the possibility of improving math performance through spatial learning depends to a large extent on whether spatial skills can be taught and learned.

Historically, spatial ability has been viewed as a core aspect of intelligence. Beginning in the early 20th century and spanning to the present day, psychologists have consistently identified spatial ability as an essential factor in the study and definition of intelligence. Perhaps owing to the close tie between spatial ability and intelligence, spatial reasoning is commonly viewed as a fixed intellectual trait – "either you have it or you don't" (Newcombe, 2010). It is not uncommon, for example, to hear someone remark that they "don't read maps," "can't follow directions," or even go as far as saying they "have no spatial sense whatsoever." This "fixed" viewpoint appears to be based on misconceptions and false belief.

Decades of research confirm that spatial reasoning is malleable and subject to improvement with practice and targeted interventions. The most conclusive evidence that spatial reasoning is malleable comes from a recent meta-analysis that analyzed 206 spatial training studies over a 25-year period (1984–2009; Uttal et al., 2013). The study concluded that people of all ages and through a wide assortment of spatial training interventions (e.g., video games, course training, spatial task training) demonstrated significant gains in spatial reasoning. Moreover, the average effect size of training was large and approached a half standard deviation (0.47). To put this effect in context, an improvement of this magnitude would approximately double the number of people with the spatial skills associated with being an engineer (see Uttal et al., 2013). Indeed, the implications of improving spatial skills are significant and far reaching, especially in relation to the ever-important STEM disciplines. Verdine et al. ("Finding the Missing Pieces," 2014) suggest that spatial instruction can be expected to have a "two-for-one" effect, yielding benefits in both spatial reasoning and mathematics.

Although the majority of studies (67%) in the meta-analysis measured spatial skills immediately after training, some studies demonstrated that the effects of training persisted over time. In one longitudinal study, the training effects were still present four months after the intervention (Feng, Spence, & Pratt, 2007). Another notable feature of the meta-analysis was the finding of nearly identical near and far transfer effects. That is, training of one spatial skill led to improvements on spatial tasks closely related to the trained skill (i.e., near transfer) as well as spatial tasks that were quite distinct from the trained skill (i.e., far transfer). For example, in two studies, mental rotation training resulted in improved mental rotation skills (i.e., near transfer), but also led to more generalized mental transformation skills, as evidenced by improvements on a mental paper folding test (Wright et al., 2008; Chu & Kita, 2011).

Overall, the results of the meta-analysis performed by Uttal et al. (2013) go against the common misconception that spatial reasoning is "fixed" and consequently "unteachable." On the contrary, spatial reasoning appears to be highly malleable. Moreover, a wide variety of training methods appear effective in bringing about durable and transferable improvements in people of all ages.

A survey of interventions and activities to support young children's spatial reasoning

In this section, we provide an overview of the types of interventions and activities that have been found to support the development of young children's spatial reasoning, including block building, puzzle play, drawing exercises, and paper-folding activities, including origami. Each one of these activities simultaneously targets a number of important spatial skills and to varying extents all encourage the development of spatial visualization skills – a feature and type of spatial reasoning that is closely linked to mathematics performance (Mix & Cheng, 2012). Together, the interventions detailed below offer an assortment of "easy-to-implement" classroom activities and ideas for lessons that provide multiple entry points to engage, support, and improve students' spatial reasoning skills.

Construction play

Construction play with materials, such as wooden blocks, Lego™, and Meccano™ toys, has been closely linked with the development of spatial reasoning (e.g., Casey, Andrews et al., 2008; Nath & Szücs, 2014). Construction play affords opportunities to develop spatial reasoning through physical and visual experiences involving the composition and decomposition of 3D structures, perspective taking (e.g., moving around one's structure), symmetry, and transformations (e.g., rotations, translations, reflections). Although interest and time spent engaging in construction play has been related to spatial reasoning (cf., Robert & Héroux, 2003; Doyle, Voyer, & Cherney, 2012), more recent research has revealed that it is the quality, both accuracy and complexity, of the building that seems most salient.

Casey, Andrews et al. (2008), for example, conducted an intervention study with kindergarten children in which they studied the effectiveness of different types of block-play on students' spatial reasoning skills. In this study, Kindergarten classrooms were assigned to one of three block-building groups. One of the groups engaged in free, unguided block play. A second group of students were given specific building goals for their block play (e.g., build a wall that could be used to contain animals). Finally, the third group were provided with these same building goals but embedded in a narrative (e.g., a story involving a dragon, Sneeze, who required help re-building a series of fallen down castles). Importantly, in all three conditions, children spent an equal amount of time engaging in block play. Both before and after the various interventions, all of the children were assessed on a number of measures including an assessment of block building complexity, 3D mental rotation, and block design – a common intelligence test that involves recreating 2D geometry designs using variously coloured and patterned cubes.

Results showed that, compared to those who engaged in free block play, children in the goal-directed block play groups demonstrated significant gains on the block design test. However, only those children in the narrative condition demonstrated significant gains in their block building performance. These findings are important as they indicate that the quality of block building influences the development of children's spatial reasoning skills. This study adds to a growing body of research that demonstrates the importance of providing children with teacher-guided block-play that involves specific building goals (e.g., Reifel & Greenfield, 1983; Hanline, Milton, & Phelps, 2001; Gregory, Kim, & Whiren, 2003).

Puzzle play

There is growing recognition that puzzles (e.g., jigsaw, Tangrams™, pentomino challenges, etc.) provide meaningful opportunities, especially in the early years, to build spatial skills. Just like construction play, puzzles target a number of different spatial skills, including composition and decomposition of shapes, mental rotation, and logic-based reasoning about spatial relationships (e.g., "this one must be a corner piece"). One recent study demonstrated just how important early puzzle play might be in contributing to the development of spatial skills (Levine et al., 2012). In this study, the authors observed child–parent(s) interactions in their homes every four months while the child was between 2 and 4 years. When the children were 4.5 years, they were assessed on a spatial task that involved mental transformations of various 2D shapes (see Figure 2.4c). This study yielded two important findings: children who were observed playing with puzzles during the visits over the 2 years performed better on the spatial task than those who did not engage in puzzle play. This relationship held even after controlling for important parental variables such as education, household income, and parental language. Moreover, the frequency and quality of puzzle play – amongst those who did play with puzzles – was further predictive of how well the children performed on the task. This study suggest that even before formal school entry certain home activities, such as puzzle play, are related to later spatial skills (Doyle et al., 2012; Robert & Héroux, 2003).

Cross-sectional studies conducted with early elementary students have further solidified a link between puzzle performance and spatial reasoning skills. For example, Verdine et al. (2008) found high correlations between performance on a standard jigsaw puzzle and measures of mental rotation, spatial perception, and spatial visualization. In an attempt to harness the relationship between puzzle play and certain geometry and spatial skills, one group of researchers designed and carried out a one-month "puzzle" intervention (Casey, Erkut, et al., 2008). Children were either assigned to a control condition (i.e., free play) or an experimental condition that involved listening and responding to a narrative. As part of the experimental condition, students worked through a series of open-ended puzzle tasks involving Tangrams. The results indicated that all boys – regardless of group assignment – demonstrated approximately equal gains on a pre- and post-test involving various puzzle tasks (designed to assess part–whole understanding). Interestingly, girls in the experimental condition but not the control group demonstrated significant gains on the task. This study suggests that, at least for young girls, puzzle play might be one effective approach for improving the early understanding of part–whole relations.

Perhaps the most promising puzzle intervention to date involves training with the video game Tetris (Okagaki & Frensch, 1994; Terlecki, Newcombe, & Little, 2008); a fast-paced puzzle game that involves rotating and translating polyomino shapes into the most optimal position. Although studies have only been carried out with adolescents and adults, the findings from these studies suggest that game play results in improved mental rotation skills and spatial visualization. There is no reason to suspect that games such as Tetris would not also be useful in early learning settings.

Drawing tasks

At some point in elementary mathematics, one must confront the often-difficult task of interpreting and creating isometric drawings – a feat that requires spatial reasoning skills. In a number of studies with engineering students, researchers have demonstrated the promising effects of training these students in perspective drawing skills. Indeed very promising research studies have shown that intensive drawing practice, of the sort that involves learning how to accurately represent 2D and 3D objects, can significantly improve engineering students' spatial skills (Baartmans & Sorby, 1996; McAuliffe, 2003; Sorby, 2009).

Recent evidence suggests that drawing activities might also be an effective way of improving young children's spatial reasoning. For example, Tzuriel and Egozi (2010) carried out a drawing intervention with a population of first grade children. The intervention consisted of eight 45-minute sessions based on Quick Draw™ stimuli (see Wheatley, 1996). Working with small groups of children, the experimenter presented students with a 2D geometric design for only three seconds (see Figure 3.4 for an example). Children then had to draw the image from memory. This was followed by a group discussion that involved sharing how the images were initially perceived and remembered (e.g., "I saw an 'X' and a 'T' overlapping one another inside a square"). The experimenter facilitated the discussion and directed children to acknowledge the different perspectives amongst group members. Children were also encouraged to rotate the images and notice how different orientations influenced one's perspective. Compared to a control group, children who participated in the Quick Draw activities demonstrated significant improvements on two separate tests of mental rotation.

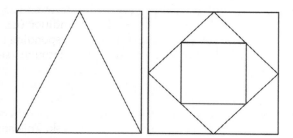

Figure 3.4: Example of the types of stimuli used in Quick Draw™. Students are presented with the image for three seconds and then must try to recreate the image from memory (for original Quick Draw™ images, see Wheatley, 1996).

Paper folding

One of the most established tests of spatial visualization is called the Paper Folding Test (see Figure 3.5). In this test, participants are presented with a sequence of folds in a piece of paper. The folded piece of paper is then punctured with a hole punch. The objective is to determine what the piece of paper would look like if unfolded – how many holes would appear and where exactly would they be located? Given that spatial visualization lies at the heart of spatial reasoning, many interventions aim to improve this important skill. Thus an ideal candidate for improving spatial visualization involves interventions that utilize paper-folding tasks.

In one such study, Taylor and Hutton (2013) led a unit on origami and paper engineering with classrooms of fourth grade students. The intervention largely consisted of creating and deconstructing complex paper models and visualizing the results of making certain folds and cuts. Compared to a control group, children in the experimental group improved on two separate tests of spatial visualization, both of which involved mental paper folding. Furthermore, participants in the program reported high levels of engagement throughout the program.

Cakmak, Isiksal, and Koc (2014) also used origami as an instructional approach to teach geometrical and spatial reasoning skills. In this study, students in Grades 4 through 6 participated in ten, 40-minute, in-class origami sessions. The teacher facilitated each session by first showing the students how to perform a certain fold and then had students follow. Throughout the instructional sequences students were encouraged to work together. After each folding stage, the class discussed the formed shapes and their properties (e.g., "Which shape do we have now?" "Why do you think we get this shape?" and "What are the properties of this shape?"). Upon completing the origami model, the teacher and students summarized the geometrical concepts and mathematical terms encountered throughout the model making. Not only did students demonstrate large gains on an extensive battery of geometry and spatial reasoning, but students also reported an increased awareness of how origami related to mathematics (e.g., geometrical transformations, fractions, 2D and 3D shapes, angles, etc.). For example, in the words of one student, "While making the samurai hat, we talked about the trapezoid, isosceles triangle, equilateral triangle, and scalene triangle. We also emphasized that the top and bottom bases of the trapezoid were parallel to each other. We folded the angles of 45° and 22.5°" (p. 65). Interestingly, other researchers have shown that paper folding offers a potentially powerful entry point into students' thinking about fractions and multiplicative reasoning (e.g., see Empson & Turner, 2006). Taken together, it appears as though paper folding is a useful tool for the teaching and learning of skills and concepts related to both spatial reasoning and mathematics.

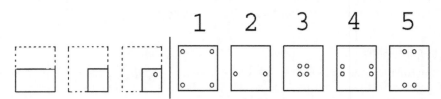

Figure 3.5: An example item from the Paper Folding Test (Chu & Kita, 2011).

Training spatial reasoning to support mathematics learning

Despite the historical relationship between spatial reasoning and mathematics, surprisingly few researchers have examined whether spatial training generalizes to mathematics learning. To our knowledge, there are only three studies that have directly tried to assess the use of spatial training for improvement in mathematics. In the next section, we review these three studies: one conducted in a laboratory setting; one in an afterschool program; and one in early years classrooms.

Training on a spatial task

In the first study to causally demonstrate the effects of spatial training on mathematics, the researchers, Cheng and Mix (2014), assigned children to either a spatial training condition (i.e., mental rotation training) or crossword puzzle condition. Both before and after the intervention, children completed two spatial tasks and a test of mixed calculation problems. Children in the spatial training condition were trained on the Children's Mental Transformation Task (see Figure 2.4c, p. 23). This training involved two steps. Children first were asked to visualize the solution to each problem. That is, to identify the correct solution amongst the four alternatives. Children then confirmed the accuracy of their response by putting together actual cardboard "cut outs" of the shapes. Thus, children were given immediate feedback about the accuracy of their mental transformations. In both conditions, the intervention lasted for a single 40-minute training session. Remarkably, children in the spatial training group, but not the crossword condition, demonstrated significant improvements not only on the mental transformation task – an expected finding – but also on the calculation test. Improvements were most evident on missing term problems (e.g., $5 + \_\_\_ = 7$), a finding that was attributed to the possibility that training primed children to approach the problems through spatially reorganizing the problems (e.g., $5 + \_\_\_ = 7$ becomes $\_\_\_ = 7 - 5$).

This is an important finding, as it is the first empirical study to demonstrate the potential of spatial training as a means to facilitate calculation performance. Other studies are needed to replicate this finding. Not all brief interventions of this sort – no matter how well they are designed – will result in improved mathematics performance. Furthermore, failure to replicate findings from such a short intervention does not necessarily indicate that spatial training does not help facilitate mathematical understanding.

Indeed, it is our belief that while carefully controlled experimental studies are necessary for moving the field forward, we also need the types of training studies that take place in actual classrooms and for sustained periods of time. In addition, while the study above targeted only one specific spatial reasoning skill, little is known about how targeting multiple spatial skills might affect mathematics learning. What follows is a description of two studies that have attempted to address some of these issues.

A construction-based afterschool arts program

Working with underserved and at-risk preschool populations, Grissmer et al. (2013) designed and carried out an extremely intensive spatial intervention. Half the preschoolers were assigned to the experimental group and half were assigned to the "business as usual" control group. Children in the experimental group took part in a seven-month intervention that was aimed at developing spatial and fine-motor skills. Four times a week, for approximately 45 minutes, children took part in activities that involved creating and copying geometric designs made from a variety of materials, including Lego™, Wikki Stix™, and pattern blocks. Although both groups of children started at the same place in terms of their testing performance, there were marked differences between the two groups at the end of the intervention. Compared to the control group, those in the spatial intervention demonstrated gains in a number of areas, including spatial reasoning, self-regulation, and importantly in overall mathematics performance. In terms of mathematics performance, children in the spatial group advanced an impressive 17 percentile points, from 30th to 47th percentile, on a nationwide test of numeracy and problem solving. This finding provides some preliminary evidence that an intensive and sustained spatial program, that utilizes a number of different spatial tasks, is an effective means of supporting young children's mathematical development. This study also points to the promising effects of interventions that aim to strengthen young children's construction skills.

A seven-month in-class spatial reasoning intervention

As part of an ongoing professional development research project, Math for Young Children (e.g., Moss et al., in press), researchers worked with a group of Junior Kindergarten (preschool) to 2nd grade teachers in three schools primarily serving First Nations populations. The researchers implemented an in-class intervention in which spatial reasoning tasks were incorporated into the regular mathematics curriculum. Teachers in both the experimental and control groups participated in separate professional development sessions. The teachers from the experimental group received professional development on teaching and learning of spatial reasoning; and teachers in the control group worked on inquiry approaches to environmental science.

The spatial reasoning intervention was delivered by the teachers over seven months and consisted mainly of a series of brief spatial tasks, which became known as "rug activities." The rug activities included drawing, building, copying, and visualization exercises (see Table 3.1) and targeted the development of the young students' intrinsic-dynamic spatial reasoning (Mix & Cheng, 2012). These activities were carried out with the full class during "circle time" or with small groups at teacher-guided math centers. On average, students participated in the activities three times a week for a total of approximately 40 hours throughout the school year. To assess the efficacy of the intervention, all of the students (N=67) participated in pre- and post-assessments of spatial language, visual-spatial geometric reasoning, 2D mental rotation, number knowledge, magnitude comparison, and a

Name of "rug activity"	Description of activity	Geometry and spatial skills targeted
1. Can you draw this? 	• Children were provided with pieces of paper with an outline of a square on it • Children were then shown a geometric design composed within the square boundaries • After viewing the design for 10 seconds, children attempted to re-create (using a pencil) the exact design within the boundaries of their own square • Teachers facilitated discussions around strategies and different ways of remembering the designs • Note: this activity was based on Wheatley (1996); also see Tzuriel & Egozi (2010) for a study on the effectiveness of this activity.	• Visual-spatial memory/ visualization • Composing/decomposing/ partitioning space • Proportional reasoning
2. Can you build this? 	• Similar procedures to "Can you draw this?" • Children were shown a geometric structure composed of multilink cubes • After viewing the structure for 10 seconds, children attempted to re-create the structure from memory using their own multilink cubes • In another version of this activity, children were presented with a structure and asked to re-create it with no memory component	• Visual-spatial memory/visualization • Composing/decomposing 3D shapes
3. Building with the Mind's Eye 	• Children were given oral instructions in how to build a 2D or 3D shape (e.g., "Take two blue cubes and attach them together, one on top of the other. Stand up the two attached cubes and make them look like a tower. Now take a red cube and attach it to ...") • Children built images of the shape in mind, based on instructions given • After giving instructions, teacher showed children multiple shapes and had children discuss/reason which one perfectly matched the description	• Visualization • Composition of 2D shapes, 3D shapes • Mental transformations • Spatial language comprehension • Visual-spatial working memory
4. Shape Transformer 	• Modeled after the "Function Machine," an "input/output" activity typically done with numbers (e.g., input = 2, output = 4; input = 5, output = 10,…etc. Rule, $y = 2x$) • In this version, input and output functions deal with spatial relationships (e.g., transformations) • Children were presented with a "machine" made out of a poster board, with "input" and "output" slots cut out • Teacher (and eventually students) prepared input and output cards to enter and exit into/out of the "machine" • Children watched and paid attention to relationship between input and output cards and tried to predict the transformation (e.g., each shape that goes into the machine gets rotated 45°)	• Mental transformations/visualization • Visual-spatial reasoning/ deductive reasoning • Composition/decomposition of 2D shapes
5. Barrier Game 	• Children worked in pairs with a barrier (folder) in between them and each with their own building materials (e.g., pattern blocks or multilink cubes) • One partner built a shape and described how to build the shape to his/her partner, who built according to the instructions provided • Children then compared their structures before reversing roles	• Spatial language • Visualization • Composing/decomposing 2D shapes/3D shapes

Table 3.1: Examples of "rug activities" carried out in the experimental classrooms.

repeated measures control task assessing receptive vocabulary. While there was no significant difference in performance between the two groups at the beginning of the school year, there were remarkable differences at the end of the year. Compared to the control group, children who had participated in the spatial activities demonstrated widespread improvements on all of the spatial measures, including spatial language, 2D mental rotation, and visual-spatial geometric reasoning. And surprisingly, a significant difference emerged on a test of symbolic (i.e., Arabic digits) magnitude comparison; a test shown to be significantly related to children's arithmetic performance (Nosworthy et al., 2013). This finding was not expected, as the intervention did not explicitly focus on number development. This is a novel finding and one that suggests the possibility that early spatial instruction not only benefits children's spatial competencies but might also contribute to the development of early numeracy skills.

On a final note, as has been reported in other spatial intervention studies (Cakmak et al., 2014), both the teachers and students reported high levels of enjoyment and engagement throughout the intervention. The teachers who led the intervention agreed that the spatial activities offered multiple entry points for their diverse learners, and furthermore, led to new insights into the potential for spatial reasoning to serve as an important foundation for mathematics learning.

Linking ideas

We are entering an exciting and promising era of spatial reasoning research. It is no longer enough to show that spatial reasoning and mathematics are related. The time has come to explain why the two are related, and furthermore, to mobilize and apply our existing knowledge in fruitful and long-lasting ways. To the latter point, we see early years education as an important place to begin such efforts. In this chapter, we shared a working hypothesis of how and why spatial reasoning and mathematics go hand-in-hand, paying particular attention to how spatial reasoning can provide an important foundation for mathematics learning. Indeed, there is now extensive evidence that spatial reasoning is malleable and can be improved in people of all ages and through a wide variety of training techniques. Although the majority of spatial training studies have been conducted in carefully controlled "lab" experiments, the educational implications of these findings are significant and potentially far-reaching. In terms of early years mathematics education, there is an accumulating body of intervention studies pointing to the importance of providing opportunities for high quality construction play (e.g., building blocks), puzzle play, drawing exercises, and paper folding. In moving forward, we urge psychologists and mathematics educators to work together in both the design and implementation of classroom-based spatial interventions.

Section 2

If spatial reasoning is so important, why has it taken so long to be noticed?

SECTION COORDINATOR: BRENT DAVIS

Chapter 4: A history and analysis of current curriculum

Contemporary school mathematics curricula tend to include few opportunities to develop, apply, and integrate spatial reasoning. This chapter explores why this might be the case by delving into the origins of modern mathematics curricula and examining some of the sources of its stability.

Chapter 5: Spatial knowing, doing, and being

The very phrase "spatial reasoning" juxtaposes reason and space, suggesting a central role of the body for reasoning. This chapter looks into the literature of embodiment, how it addresses mathematics reasoning, and in particular, spatial reasoning.

4

A history and analysis of current curriculum

BRENT DAVIS, MICHELLE DREFS, KRISTA FRANCIS

In brief ...

Contemporary school mathematics curricula tend to include few opportunities to develop, apply, and integrate spatial reasoning. This chapter explores why this might be the case by delving into the origins of modern mathematics curricula and examining some of the sources of its stability.

Where does mathematics curriculum come from?

Since the commencement of standardized education in the 18th century, most formal curricula have been framed in terms of meeting needs. The matter of whose needs has always been a point of contention, with the poles of such debates having tended to focus on the specific needs of individuals and the general(ist) needs of society.

Given the cultural and economic conditions 300 years ago, at the dawn of modern public schooling, it is easy to see why arithmetic and algebra were so heavily emphasized in the curriculum. Rapid industrialization, extensive urbanization, and a new scientific mindset presented needs for a particular sort of numeric literacy. However, in the intervening centuries a gap has opened between the foci of school mathematics and the nature of the mathematics actually engaged by most in their day-to-day lives. This point is particularly obvious around spatial reasoning – which, as detailed in Chapter 1, is scarcely addressed in schools even while it is of mounting relevance in contemporary society.

The question of where school mathematics curricula come from is not as easy to answer as one might think. Many of the key influences are so tangled and obscured that any attempt to offer a comprehensive account of beginnings and evolutions would inevitably drift toward a fiction. The intention of this chapter is thus more to give a sense of the complex history of school mathematics curricula, focusing in particular on those programs of study that are typical in the English-speaking world. Our strategy to keep the discussion manageable is to organize it into three periods that might be roughly characterized as pre-industrialization, early industrialization, and post-industrialization.

Our intention here is not to provide a detailed account of the emergence of mathematics curriculum. For that we defer to the work of others (Schubring, n.d.; Howson, 1973; Bishop et al., 1996; Stanic & Kilpatrick, 2003; Menghini et al., 2008; Sinclair, 2008) and proceed with a broad-brush-strokes account developed around a handful of defining moments, starting with a shift in sensibility that occasioned the emphasis on practical and procedural mathematics content.

Pre-industrialization – prior to the 1700s

Western society underwent a major epistemological shift several hundred years ago. Having been smoldering for many centuries, the major change in cultural sensibility caught fire over in the 1600s with the Scientific Revolution – and, later, amplified in the late 1700s with the Industrial Revolution. The impacts of these events on formal education were immense, and they are readily identified in news stories, popular writings, and art of the time. An example is illustrated in Figure 4.1. The image on the left, titled *The School Master*, dates from the mid-1600s. It portrays a teacher in the center of a group of children of different ages and sexes who are engaged in activities that appear to range from the playful to the academic. The untitled image on the right is much different. Here the teacher in an English school is on the side of the room, a position that enables surveillance of every student in the homogeneous grouping as they study the same thing at the same time and work toward similar levels of expertise. A completely different picture of schooling is presented.

Why the shift? We suspect that it has much to do with assumptions about knowing and learning. To elaborate, through the late 1500s to the

Figure 4.1: A pair of images that illustrates a transition in schooling in the 1600s.
LEFT: *Le maître d'école* (Adriaen van Ostade, c. 1650).
RIGHT: Untitled (Artist unknown, c. 1750).

late 1600s, European societies underwent what some philosophers call "the Epistemological Turn" (cf., Hiley & Bohman, 1992), a transition from one way of thinking about knowledge to another. This point is actually a difficult one to make in English – which, in spite of having the largest vocabulary of any language, seems to have a poverty of expressions for "knowledge." Most other European tongues have a greater variety of terms to refer to different ways of knowing. One example is the ancient Greek pairing of *gnosis* and *episteme*.

Gnosis refers to a category of knowledge that is about meaning, relationship, wisdom, and purpose. The development of gnosis is a life-long process, and it is associated with artistic and poetic genres. One cannot reduce gnosis to bite-sized, learnable bits; rather, the preferred pedagogical strategy is through myth, allegory, parable, metaphor and other associative means to link categories of experience, moral import, and ethical weight. When schools (i.e., formal settings to collect individuals with the intention of developing knowledge) were first invented by the ancient Greeks, it was this category of knowledge that they were designed to address. Indeed, this sensibility is embedded in the word *education* – derived from roots meaning "to draw out" (i.e., the deep knowledge of the universe that was deemed to be knitted into one's being). The concern with gnosis also aligns with formal education's original focus on the "liberal arts" – literally, arts that are freeing. The sensibility is well portrayed in such works as Raphael's *The School of Athens*. It is also hinted in the original meaning of the word *school*, which is derived from the Latin *schola*, "intermission of work, leisure for learning; learned conversation, debate."

By contrast (or, perhaps, in complement), *episteme* refers to facts and practical knowledge. While it was never deemed unimportant, it was originally assumed to comprise those things that you would learn as you needed to learn them. That is, epistemic knowledge was not originally seen as appropriate to a formal education. The stuff of everyday know-how, of experiment, of analysis, and so on would take care of itself as necessary. The modern turn toward this category of knowledge is reflected in contemporary understandings of the word *school*, which pull away from its heady origins as a place of learned conversation and press more toward factory-like sites for the perpetuation of established truths.

Before delving into the significance for conceptions of schooling of these two categories of knowledge, it is worth pausing for a moment to reflect on the manner in which mathematics can be rightly viewed as an instance of both gnosis and episteme. A cursory overview of the etymology of the word *mathematics* reveals related expressions in other languages denoting notions such as "wide awake," "wise, sage," and so there is good historical reason for the manner in which mathematics has come to span gnosis ("to care," "wide-awake," "lively," etc.) and episteme ("skills," "a lesson"). It is a tension that in a very real sense continues to be lived out in universities in the ongoing debate on whether departments of mathematics belong in faculties of arts or faculties of science. More recently, the issue has reasserted itself in the rocky emergence of an empirical mathematics that is motivated as much by experiment as it is by creative associations.

Within the institution of the school, however, mathematics is rarely conceived in terms of spanning types of knowledge. Rather, across varied enactments of formal education, mathematics has been cast almost exclusively as one or the other. The shift in emphasis from "mathematics as an instance of gnosis" toward "mathematics as an instance of episteme" occurred with the invention of the modern school during the Scientific and Industrial revolutions. Let us attempt to unpack this assertion by highlighting five moments in the emergence of mathematics curriculum, clustered around five personalities. We do not intend to suggest these personalities are responsible for the evolutions described. Rather, they are offered more as familiar and historical markers.

We start with Pythagoras (575–490 BCE). He and his school of geometers are often credited with many innovations to schooling, both structural and curricular, but their major contribution was conceptual/philosophical. Pythagoras helped to collect a rather disjoint set of facts and insights into a coherent, powerful system of knowledge. Phrased somewhat differently, his major contribution to the institution of education revolved around a conversion of practical know-how (episteme) into theoretical knowledge (gnosis). The Pythagoreans recorded, developed, and delved into concepts *for the sake of those concepts*, not for application, communication, pedagogy, or other purposes. In the process, a sort of geometry curriculum began to come to form – although its curriculum structure was quite dissimilar to the modern school's linear trajectory through a parsed subject matter.

By Plato's time (424–348 BCE), formal education was a common feature among the citizens of Greece. With its rise in popularity, along with ever-growing cultural knowledge (i.e., of the gnosis sort), came the need for greater structure. One of Plato's major contributions was the subdivision of Liberal Arts into the Trivium (Grammar, Logic, and Rhetoric) and the Quadrivium (Arithmetic, Geometry, Music, and Astronomy). This particular disciplinization of knowledge persisted into Medieval schools, and the attitude of slicing knowledge into disciplines has been a hallmark of the scholarly world ever since.

Importantly, however, "mathematics" was not included in Plato's construct, for a simple reason. Mathematics did not yet exist as a coherent disciplinary domain. The devices had not yet invented to unite Logic, Arithmetic, Geometry, and other domains, but Euclid (323–283 BCE) took a major step in that direction as he used logic to prove, connect, and extend geometric truths. At the same time, he sequenced key ideas – for example, placing planar geometry prior to solid geometry. These intellectual leaps set the stage for a unified, well-structured discipline. (As we will see, it also had an immense impact on the content of modern curricula.)

The unification of mathematics into a coherent discipline had to wait almost two millennia until René Descartes (1596–1650 CE) brought together arithmetic, analysis, geometry, and logic through the masterstroke of linking number and shape through a coordinate system. That extraordinary intellectual contribution marked the emergence of the system of knowledge that we know as *Mathematics*, affording a means to collect together not just knowledge

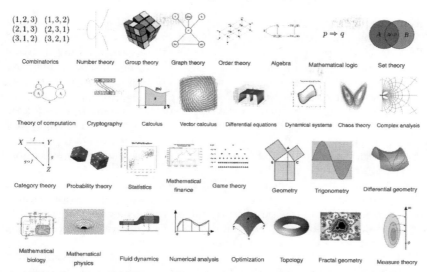

Figure 4.2: Descartes' masterstrokes of (1) uniting number and shape through a coordinate system and (2) insisting that all claims to mathematical truth be grounded in logical argument set the stage for the emergence of the discipline of mathematics that now comprises many other branches of inquiry. As reflected in the iconic images above, the intertwining of quantity and shape is foundational to most branches of mathematics. (The list of branches of mathematical inquiry and the associated images were drawn from the Wikipedia.com entry, "Mathematics" on April 30, 2013.)

of number, shape, and argument, but also a host of other foci that we now understand as properly mathematical. This shift was signaled linguistically with a move from the singular *mathematic* to the plural *mathematics* (as noted in the above etymology; see also Figure 4.2). It has since come to serve as a defining aspect of the Epistemological Turn in the world of philosophy.

To explain, while Descartes contributed a great deal to what was to become the school subject of mathematics, he did very little to shape mathematics curriculum. However, others had already begun that work. Many personalities of the era worked to develop ways to take advantage of the new technology of the printing press – which itself might be argued to be a major contributor to the very possibility of a *standardized* curriculum. Of these personalities, Robert Recorde (1512–1558 CE) of England stands out. He published the first English mathematics textbook in 1540, and in it he made the important move from a text based on prose to a text based on a standardized notation. Like many of its precursors, including the notable *Liber Abaci* from Leonardo of Pisa in the 1200s, Recorde's text was decidedly algorithmic and utilitarian in character. Further, and unlike its precursors, Recorde's text was mass produced and ready made for an emerging linearized curriculum mindset.

With a standardized resource in place, alongside society-wide movements to bring children into state-funded schools and a uniformity-of-production mindset, the necessary conditions were present for a shift in definition of curriculum. Prior to the 1600s, curriculum was understood as it is in the phrase "curriculum vitae"; it was a path-based metaphor that was useful for describing one's movement through existence, usually post facto and with a decidedly gnosis flavor. By the 1700s, the dominant interpretation of curriculum was moving toward a standardized, age-indexed trajectory of study. It was still a path-based metaphor, but one that was prescribed in advance and with a clearly epistemic focus.

Notably, the work of curriculum developers of the time (and since) reflected some deep-seated beliefs about the nature of mathematics knowledge. Two assumptions in particular figured into the emergent contents and structures of school mathematics. Firstly, mathematics curricula were structured linearly and logically. On this count, curriculum developers drew explicitly on the model of the Euclidean proof, according to which one began with basics/fundamentals and moved systematically toward more sophisticated constructs. Secondly, mathematics curricula were selected and organized to recapitulate the history of mathematics, starting with number, developing binary operations, and culminating in algebra and analysis.

Many modern educators would see these as instances of sound and sensible practice – a reading that betrays more a deep-seated cultural bias than any great insight. In fact, we are aware of no research into learning that demonstrates the "fitness" of the curriculum structures and contents we have inherited. There is, however, a growing body of commentary and research that renders the above assumptions problematic. Humans, for example, are not logical creatures, but association-making beings whose capacity for logic rides atop irrepressible tendencies to see connections and make intuitive leaps. Further, as reflected in Figure 4.2, the contents of curriculum that were selected in the 1600s (and that have changed surprisingly little in the intervening centuries) currently represent a very impoverished view of the field of mathematics.

Early industrialization – 1650 to 1950

Educators often describe their experiences with curriculum revisions and innovations in terms of a swinging pendulum, pointing to what feels like the endless oscillations between "skills and understanding" (or, depending on the audience, "applied and pure math," "procedural/rote competence and conceptual understanding," or "back-to-basics and problem solving").

We would argue that what teachers experience as an endless oscillation may be more a residue of an unresolved conflict. There was a forced marriage of two different sorts in school mathematics about a century ago, and Western society is still dealing with the fall-out. What is experienced as an inevitable mood swing is actually a case of a split personality.

In the early 20th century, there were significant differences between mathematics curricula at the elementary and secondary levels. Secondary schools of the pre-1900s tended to be somewhat closed institutions. They were intended for various elites – in particular, those from higher economic classes and those of "rare aptitude." These schools operated with an assumption that they were educating the "leaders of tomorrow" – persons who need not be burdened with meaningless procedures and extensive drills necessary to develop flawless mastery. However, they did require schooling in reasoning and in great ideas across the knowledge domains deemed to have greatest cultural value. These desires to cultivate reason and to prepare an even more rarefied elite for tertiary education helps to explain why secondary mathematics looks as it does: a preparation for calculus, with mind-cultivating side trips into proof and other topics.

Economic and ideological shifts transformed the situation, however. The co-entangled emergence of a wealthy middle class and the spread of democratic ideals helped to give rise to an "education for all" movement that sought to make high school the educational standard. That movement had succeeded in most English-speaking nations by the middle of the 20th century. With that evolution, problems of coherence between elementary and secondary levels arose. For example, the industry-prescribed, calculation-heavy "arithmetic" of the early years was forced to align to the elite-serving, concept-rich "algebra" of the secondary school. A similar issue arose around the topic of geometry. As Sinclair (2008) noted, it was not until 1892 that there was any attempt to standardize geometry curriculum in the United States. To that point, elementary schools had tended to focus on utilitarian processes involving basic shapes (i.e., naming, measuring, classifying), whereas curriculum at higher levels often drew directly on Euclid's *Elements* (at least the first six books, focused on planar geometry).

Another noteworthy development also unfolded at roughly the same time as the "education for all" movement. In the late 1800s, behaviorism began to flow over the edges of psychology into schools – and, specifically, into elementary schools where its principles were readily fitted to incrementally structured curricula and emphases on rote mastery. The behaviorist emphasis on observable and measurable performance amplified the skills-based and rote-mastery foci of the elementary school as they contributed to an even further erosion of the already-limited geometric/spatial content of the school's curriculum. The impact of behaviorism on the more reason-driven secondary school, by contrast, was much less pronounced.

In brief, then, the linkage between elementary and secondary school mathematics has never been smooth. And, as mentioned above, ongoing efforts to reconcile the incoherences have been experienced by teachers not for what they are, but as pendulum swings. Indeed, there remains one rather significant difference between elementary school math and secondary school math. At the lower level, every concept has a ready interpretation or application in the everyday world. That is, the mathematics is heavily analogical, but the metaphors, exemplars, images, and other associations that are invoked are tied to everyday actions and experiences. The secondary level is also heavily analogical, but the analogies are for the most part to other mathematical concepts, creating entirely new levels of abstraction. For instance, whereas "equals" might mean "makes" in 2nd-grade counting activities or "arrives at" in 4th-grade number-line work, at the secondary level the concept is a more complex blend of "makes," "arrives at," "is the same as," "balances with," "can also be expressed as," "has the same magnitude as," "maps onto," and so on. Students must be able to move flexibly and fluidly among interpretations, selecting the one(s) most appropriate to the task at hand.

For we math educators, this discontinuity between elementary and secondary is often encountered as the cocktail-party confession, "I was good at math until Grade 6." Such remarks, for us, are not unlike the statement, "I was a confident swimmer until my feet stopped touching the bottom."

On this topic, of worthy note is a not-particularly-well-known study begun in the late 1920s that powerfully demonstrated the relationship between elementary and secondary mathematics. Louis Bénézet (1935a, 1935b, 1936), a school administrator, implemented an experiment in Manchester, New Hampshire whereby children were given no formal education in mathematics until Grade 6. What may be surprising to readers today is that it made no difference to their subsequent mathematics learning. Placed into historical context, however, the result was fully predictable. The projects of elementary and secondary education at the time were simply not aligned.

An awareness of the history of mathematics curriculum is very useful to understanding what's really behind both the pervasiveness of the cocktail-party confession and the "success" of the Manchester experiment. In many ways, we continue to act out a centuries-old history. The simple and unfortunate truth is that the elementary and secondary split in mathematics is far from resolved, and we continue to live the tensions and frustrations of a deep incoherence that is still thoroughly embodied in contemporary mathematics curricula.

Post-industrialization – 1950s onward

Over the past 70 years there has been a series of distinct movements in North American school mathematics curriculum, the more prominent and sustained of which are summarized in Table 4.1. The emphasis in these movements was not so much on the content of the curriculum. Rather the emphasis was more about the manner in which the curriculum was to be engaged.

Interestingly, each movement has been popularly seen as a reaction to world events beyond North America (indicated in the right column), but such readings are likely inaccurate or over-simplified. The fact of the matter is that the New Math was already in the planning stages before Sputnik went into orbit, more rooted in the 20th-century rise of Formalist philosophy than the Cold War. Similar can be said of the other two movements. The convenience of an enemy has been argued by some (e.g., Grumet, 1988) to be nothing more than an attempt by governing groups to deflect blame and responsibility for their poor economic decisions onto teachers and students.

However they might be characterized, there is no disputing that curriculum reform projects have been a regular feature on the mathematics curriculum landscape since the 1950s. As noted above, they have been experienced by many educators as pendulum swings – but the inadequacy of that image is revealed in the fact that they have *all* been directed at shifting the focus of instruction from calculation toward comprehension. Strategies and emphases have varied dramatically, but the goals have been stable, as

ERA	MOVEMENT	ADVERSARY
1960s–1970s	New Math – based in set theory; abstract and pure, aiming at comprehension (over calculation)	Soviet military and space technology
1980s–1990s	Standards Math – focused on problem solving, based in "real life," and utilizing emergent technologies and manipulatives; practical and applied, but still aiming at comprehension (over calculation)	Japanese cars and electronics
2000s	Reform Math – rooted in inquiry; avoiding roteness and nurturing innovative thinking, but still aiming at comprehension (over calculation)	Chinese production capacity

Table 4.1: Trends in school mathematics curriculum since the 1950s.

have the actual contents of mathematics curriculum for the most part. These stabilities are likely linked to the ongoing desire to overcome the incoherence of the unification of elementary and secondary curriculum in the first half of the 1900s. With that stability, children are compelled to master competencies that are increasingly (if not completely) irrelevant as many mathematical competencies that have emerged as necessary are ignored.

Another common feature of these movements has been a persistent insistence that greater attention be paid to the development and use of spatial-based competencies, bolstered by increasingly prominent cultural narratives of the power of image-based reasoning (e.g., Einstein's thought experiments). Calls for change were articulated initially as a recommendation for more geometry, later as an insistence for more visualization, and most recently as a reminder of the role of spatial reasoning in the development of mathematical understanding. The 1980s are particularly noteworthy in this regard, as mathematics education researchers such as Bishop (1980, 1983, 1986, 1988a), Clements (1982), and Presmeg (1986) helped to re-ignite the conversation.

Not surprisingly, just as the other moments were motivated by shifts in sensibility around knowledge and learning, this one was spurred by the rise of constructivism as a dominant discourse in the field. In particular, constructivists' attendance to bodily action and situated experience helped to underscore the importance of spatiality (see Chapter 5).

What is holding the curriculum in place?

This book challenges the assumed sufficiency of current mathematics curriculum, for reasons developed in other chapters. In this section we examine major cultural forces that hold curriculum in place.

One major force is an ongoing collective assumption that 18th-century curriculum designers "got it right." That belief is easily challenged by the current landscape of careers. Consider for example, the *Wall Street Journal*'s annual list of the 200 best and worst careers, ranked according to physical

Rank	Top 10 Careers	Rank	Bottom 10 Careers
1	Mathematician	191	Corrections Officer
2	Tenured University Professor	192	Firefighter
3	Statistician	193	Garbage Collector
4	Actuary	194	Flight Attendant
5	Audiologist	195	Head Cook
6	Dental Hygienist	196	Broadcaster
7	Software Engineer	197	Taxi Driver
8	Computer Systems Analyst	198	Enlisted Military Personnel
9	Occupational Therapist	199	Newspaper Reporter
10	Speech Pathologist	200	Lumberjack

Table 4.2: Top and bottom 10 jobs in a ranking of midlevel-income careers, assembled by Careercast.com and published by the *Wall Street Journal*.

demands, work environment, income, stress, and hiring outlook. Table 4.2 presents the top and bottom 10 from the 2014 ranking.

There are three details that we would highlight here. Firstly, most of the top careers arose only in the last century (a point that is particularly apparent in the complete listing). Schools are now tasked with preparing children for roles that have not been around long – and, indeed, for many that have not yet been imagined. Secondly, comparing the top to the bottom of the list, while it is clear that the needs for mathematical competence are considerably higher at the more desirable end of this ranking, it is just as obvious that contemporary school mathematics curricula are much better fitted to the careers at the less desirable end. And thirdly, as we consider possible gaps in curriculum content for recently developed and more desirable careers, spatial reasoning emerges as a major priority. It is central to the thought processes, the technical interfaces, and the communicational strategies of those roles.

Given that backdrop, one might expect there to be pervasive movements to update school mathematics. Why isn't that happening?

One obvious reason is the top-down structure of most curriculum development projects. To get a sense of these structures, consider the following statement:

> The Common Curriculum Framework for … Kindergarten to Grade 12, was developed by the ministries of Education in Alberta, British Columbia, Manitoba, Saskatchewan, the Northwest Territories, and the Yukon Territory, *in cooperation with teachers and other educators* from these provinces and territories. *Reaction panels composed of teachers, administrators, parents, post-secondary educators, business representatives, and members of community organizations* made important contributions.[1] (The Western and Northern Canadian Protocol, n.d.; emphasis added)

Such statements are typical. In brief, ministries of Education "develop" curriculum, educators "cooperate," and other stakeholder groups "react."

Overwhelmingly, as the above statement makes clear, curricula are "developed by the ministries of Education." It's a political endeavor, not an educational one.

This point is echoed in a similar statement about the Common Core Standards Initiative in the United States (n.d.), drawn from their website:

> The Common Core State Standards Initiative is a U.S. education initiative that seeks to bring diverse state curricula into alignment with each other by following the principles of standards-based education reform. The initiative is sponsored by the National Governors Association and the Council of Chief State School Officers.

There are two details that are worth emphasizing here, as not only are the forces behind the initiative identified, so is the principal motivation. Like the WNCP preamble, this one makes it clear that the Common Core is a government project. But unlike the WNCP preamble, this one highlights the driving force of "standards-based education reform." A bit later on the same web page, the following is offered:

> The SBE [standards-based education] reform movement calls for clear, measurable standards for all school students. Rather than norm-referenced rankings, a standards-based system measures each student against the concrete standard.[2]

The above statement is phrased in a manner that obfuscates the agencies behind the movement. SBE reform almost sounds as though it's at worst benign, and more likely beneficial. It's commonsensical, and apparently it's something that the general populace wants. In fact, however, SBE is a politician-driven movement that finds its most focused activity in international comparison tests such as the Programme for International Student Assessment (PISA) and the Trends in International Mathematics and Science Study (TIMSS). More critically, the phrase "measures each student against the concrete standard" is one that demands special attention. What is *the* concrete standard? Who sets it, and where does it come from?

In fact that standard is rooted in and derived from *the* mathematics curriculum. In other words, government agencies responsible for mathematics curricula appear to be caught up in a closed loop. One can gain precious little understanding about where mathematics programs of study come from by looking at the processes for their development and implementation. This particular point was underscored by an insight from Walter Whiteley:

> When the last Ontario Curriculum was being written, the ministry referred to "research based" – when they actually meant they had compared their sequence of topics etc. with other curricula on a national and international scale. A curricular "regression to the mean" type of phenomenon. Politicians see a risk in being an outlier. (personal communication, 2013)

Ontario isn't alone in this curious usage of "research based." The notion is little different from the invocation of "the concrete standard" mentioned in U.S. Common Core documents. Other countries, such as England, are much more overt in their motivations by, for example, commissioning reports comparing their curriculum to those of other high performing countries (Ruddick & Sainsbury, 2008; Hodgen et al., 2010). In other words, the key factor in making curriculum decisions is not utility, relevance, or facticity, but what everyone else is doing.

It may come as no surprise, then, that there tends to be very little difference among school mathematics programs of study from one country to the next with regard to the concepts and skills that populate lists of outcomes and objectives. At the moment, across jurisdictions mathematics curricula are generally organized around a central hub of mathematical competencies (e.g., reasoning, applying strategies, and communicating), a surrounding cluster of content strands (e.g., number, measurement, space and geometry, patterns and algebra, data management), and a further layer comprising specific learning goals.

As similar as the lists of outcomes and objectives might be, however, it would be wrong to leave this part of the discussion without mentioning that there are also some very big differences across jurisdictions. In Russia, for example, primary children are introduced to set theory and pre-algebra as they grapple with varied continuous instantiations of number – much in contrast to the narrow North American focus on discrete interpretations and concrete instantiations of number within a well-demarcated arithmetic. Similarly, the bugaboo topic of "fractions" is handled very differently across nations. Whereas North American children begin formal study in primary school, French and Finnish students don't see fractions until the equivalent of our high school – and then only as a special instance of rational expressions. Along different lines, New Zealand has incorporated statistical inference across grade levels, responding to the emergent need for critical competencies in making sense of data and claims based on those data.

Ironically, across all of these curriculum-development projects, fidelity to the field of mathematics does not seem to be a consideration. Whereas many of the subjects studied in school very much resemble their parent disciplines, mathematics scarcely does. In particular, it lacks the vibrancy, the connectedness, and the expansiveness of mathematics. So, in historical terms, some of the important questions emerge around when and why school math lost touch with its parent discipline, freezing into an unchanging subject matter that is not attentive or responsible to the greater world. The suggestion that spatial reasoning be included as a topic in school mathematics illustrates this point. Other topics of increased currency include complex modeling, fractal geometry, exponentiation and logarithms, computer programming, and network theory.

To this list of factors that have held the curriculum in place, it is worth mentioning the influence of work related to mathematics learning disabilities. Involving various fields of study (i.e., cognitive psychology, neuroscience, special education), the primary concern here is with identifying and assisting students who struggle to learn mathematics. Typically the intersection between learning disabilities and the curriculum is limited to discussion of the appropriateness of and required modification to the curriculum and teaching approaches (e.g., explicit instruction, graduated instructional sequence). Examining this link more broadly, however, the relatively restrictive focus within mathematics learning disability on "number" could be seen as supporting the relative inertia of the mathematics curriculum. To unpack this idea further, it is helpful to discuss both practice and research as it relates to mathematics learning disabilities and more broadly mathematical cognition.

The assessment of a mathematics learning disability (MLD) is commonly first initiated by teachers or parents who identify a student as consistently struggling to master curricular objectives. From this global approach, formal assessment of an MLD more singularly focuses on sub-average performance in one or more areas of calculation, math reasoning (problem solving), and math fact memorization. More recent diagnostic criteria also include impairment in number sense (American Psychiatric Association, 2013). A student's current level of skills and abilities in broader curricular areas (e.g., algebra, geometry, measurement, and data analysis and probability) is of limited concern in the MLD assessment; although such areas may, or may not, be examined for purposes of informing instructional practices.

A similarly restrictive range of focus is mirrored in the research informing our understanding of MLD and, in particular, the underlying processes that impacts math achievement. Over the past decade, researchers interested in the basic cognitive and neural mechanisms that undergird MLD have tended to almost exclusively focus on numerical processing (i.e., how number, both non-symbolic and symbolic, is "represented" in our brains). This focus on a single core deficit has most recently been characterized by Fias, Menon, and Szücs (2013) as a "strong bias" to the exclusion of the investigation of other critical cognitive functions and processes (p. 23). Review of topics covered within recently published studies serves to highlight this exclusionary focus, with Figure 4.3 (next page) depicting the disproportionate attention given to number (and somewhat to working memory) as opposed to other cognitive factors as contributing to mathematical difficulties.[3]

A second issue is that researchers tend to examine mathematical achievement using tasks heavily weighted toward the number domain (e.g., counting, calculation, problem solving). It follows that while researchers in this area have used the terms "mathematical cognition" and "numerical cognition" fairly interchangeably to describe their area(s) of study, the term numerical cognition is seemingly more apropos given the restricted range of cognitive processes and mathematical tasks that have been used in the study of both mathematical abilities and disabilities.

As with the broader history of mathematics previously outlined, the study of mathematics learning disabilities has not always had a primary preoccupation with numbers. Rather we see in the origins of this work, dating back to the early 1900s through case studies of patients with focal brain injuries, a somewhat wider scope of interest. As Girelli, Semenza, and Delazer (2004) outlined, spatial abilities were often considered within these early investigations as central to understanding impairments in number processing and calculation skills. While interest in spatial abilities has certainly been a through-line in the study of MLD, until more recently the topic has obtained

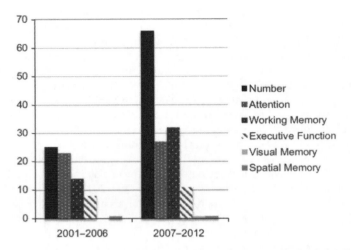

Figure 4.3: Comparison of number of studies on MLD (between 2007–2012) that have selected number as area of interest versus other cognitive correlates; totals per variable category are not mutually exclusive as multiple variables may be included in single publication.[4]

insufficient traction or development in comparison to number-focused aspects of MLD. A good example of this is David Geary's (2004) proposal of three subtypes of mathematics learning disabilities: procedural, semantic memory, and visuospatial. Of these, the visuospatial subtype has been the less studied and, not surprisingly, the results obtained inconsistent and much less persuasive.

Consequently, both practice and research related to MLD have been relatively restrictive. At minimum, this could reflect the limited ways school math is lived and perceived. It could also be argued that the restrictive approach to MLD has, in turn, resulted in little impetus or data by which to drive curricular changes and, as such, has been complicit in its support for the current curricular status quo. For example, consider what the incident rate would be for MLD should the curriculum be more broadly envisioned. Would it stay the same or possibly reduce as students are exposed to different ways to approach and think about mathematics?

So far we have identified five factors that hold curriculum in place: popular assumption about the role of mathematics, top-down structures of curriculum development, culture of examination, the separation of school math from mathematics research, and restricted practice and research focus related to mathematic learning disabilities. Other forces include the textbook/ resource industry, university-level curriculum and pedagogy, and teachers' disciplinary knowledge of mathematics. Perhaps the greatest source of cultural inertia is that, for most, "math" is what was experienced in school. That is for the vast majority "math = calculation."

The frustrating detail in all this is not so much that school mathematics curriculum has built up a powerful resistance to change. It is, rather, that even if a substantially different, temporally appropriate, culturally relevant, and learner-adaptive curriculum were to be developed, it would stand little chance of surviving in the current educational ecosystem. Politicians wouldn't be able to compare children's performances, publishers wouldn't be able to cater to the status quo, universities wouldn't know how to gate-keep, and many teachers would be at sea with what it means to think mathematically.

Linking ideas

Perhaps the forces noted above should be neither terribly surprising nor terribly dismaying. If substantial and relevant curriculum change is going to happen, it will likely be tethered to erosions of the intertwined sensibilities that hold modern school math in place. Curriculum is a result, not an input. It cannot serve as a mechanism to effect change; it must co-evolve with shifts in belief and expectation. A useful device in effecting such change is a broader awareness across stakeholder groups of where mathematics curriculum comes from and what "curriculum" is.

Curriculum is popularly assumed as an accumulation of valuable knowledge bits that are parsed out over 12 years. As noted, the model that underlies this conception is actually derived from the field of mathematics – most particularly, Euclid's geometry – where the notion of moving incrementally from self-evident fundaments to more sophisticated insights is commonplace. This conflation of curriculum structure with logical arguments may actually be an error. As Sinclair (2008) explained, the progression of ideas presented in *The Elements* was a logical and sequential structure. It was not a pedagogical one, as was assumed by many of Euclid's followers and as has been instantiated in modern curriculum.

To be clear, the prevailing image of curriculum is complicit in the entrenchment of curriculum. Other spatial models are possible – for example, ones based in other geometries, and ones that lend themselves to competencies associated with spatial reasoning.

To that end, we close this chapter by observing once again that each of the major transitional moments in school mathematics was, firstly, triggered by significant socio-economic shifts and, secondly, enabled by new ways of thinking about knowledge and learning. We believe that both these conditions are currently being met in Western societies. On the former, few would dispute the suggestion that current socio-economic evolutions are on a par with those that accompanied the Industrial Revolution. On the latter, and as is explored in Chapter 6, theories of embodiment are emerging from cognitive science and other domains that challenge current conceptions of thinking and offer new insights into mathematical knowing and learning.

Notes

1   Available through: http://www.wncp.ca/

2   Available through: http://www.corestandards.org/

3   Different models and theories exist as to the structure of and relationship between various cognitive processes. While some researchers parcel working memory as distinct from spatial (memory) abilities, others identify greater areas of overlap. One example of the latter is Baddeley's proposition that working memory consists of two temporary storage systems: a phonological loop and a visual-spatial sketchpad (e.g., Baddeley & Hitch, 1974; Baddeley & Logie, 1999). Researchers adopting this framework have found visual-spatial working memory to be associated with specific deficits in mathematical performance among children with MLD (e.g., Geary et al., 2007). As such, there may be areas of overlap between the variable categories presented in Figure 4.3.

4   Used and adapted with permission from Mazzocco & Räsänen (2013).

5

Spatial knowing, doing, and being

JENNIFER S. THOM,[1] LISSA D'AMOUR, PAULINO PRECIADO, BRENT DAVIS

─── In brief ... ───

The very phrase "spatial reasoning" juxtaposes reason and space, suggesting a central role of the body for reasoning. This chapter looks into the literature of embodiment, how it addresses mathematics reasoning, and in particular, spatial reasoning.

Interrupting Cartesian dualism

Campbell, an educational neuroscientist, has broadly summarized the deep implications of reconceptualizing mind as embodied – that is, as "a bona fide property of matter" (2011, p. 9). As he explained, an immediate entailment of an embodied theory of mind is "that any changes in subjective experience must in principle manifest objectively in some manner as changes in brain, body, and behavior, and vice versa" (Campbell, 2011, p. 10).

In this chapter we explore implications, for school mathematics, of theories of embodiment. We argue that spatial reasoning compels renewed attention to the primacy of the experiencing body. Indeed, while such theories are often regarded as "radical" or even "fringe," the topic of spatial reasoning helps us to recognize that embodied perspectives on knowing and learning actually represent a powerful site of convergence for multiple, mutually reinforcing fields of inquiry. In particular, a half-century of cognitive science research has come to foreground the knowing body in human cognition.

At the same time, school mathematics continues to be principally oriented by perspectives on cognition that focus on abstract mental operations. These perspectives often minimize, and occasionally dismiss, the primacy of embodied interactions in shaping not only personal understandings, but also the person – the thinking self, the acting agent. Schooling practices, as described in Chapter 4, remain mired in a mind–body split that divorces a supposed transcendent mind from a base, discountable, unrefined, bodily being.

The phrase "Cartesian dualism" (or Cartesian mind–body split or dichotomy) is often used to refer to this separation. In this chapter, we elaborate on embodiment as an alternative to Cartesian dualism. Extending the previous chapter's discussion of the emergence of contemporary school mathematics, here we present, develop, and draw from embodied theories of cognition that together point to spatial reasoning as vital and central to the project of moving school mathematics curricula out of its Industrial Era rut and into a rich future.

Roll, slide, and stack: a vignette from the classroom

We frame the ideas and discussions in this chapter with a vignette of a class taught by Jennifer, one of the authors. This episode is from a larger study of a Grade 2 class of 23 children in which the students learned about 3D and 2D geometry through partner, small-group, and whole-class explorations. As the chapter unfolds, we take the reader through paradigmatically different historical interpretations of this learning moment in geometry. Our focus is on how these interpretations position the role of the body – including its movement in space, the experience and recursive re-membering of such past and present movements, and the dynamic interplay of acting and perceiving – in the development of spatial reasoning.

Jennifer is teaching a class of grade two students. To date, their studies have focused on 3D and 2D geometry. While talking about what objects can do, Jennifer notices that despite the fact that the children use the same words – "roll," "slide," and "stack" in their conversation about objects, the meanings that individual students have for these terms may not be obvious to the whole class. In an attempt to enable the children to make sense of these concepts as a group, Jennifer shifts the topic of the conversation and the class's attention to articulating what it means to roll, slide, and stack.

> Jennifer asks the whole group, "What does it mean to roll?"
>
> Jade answers the question moving an outstretched arm in a circular motion three times. "So if you don't have any sides and you start rolling in circles."
>
> Jennifer looks at her and asks, "Can you think of something that rolls?"
>
> Jade thinks for a second and responds, "Like that ball," walking up to the front table and pointing to the sphere that was there, along with other 3D objects.
>
> "So, you are saying that this 'ball' will roll, OK? What do you mean by roll?" replies Jennifer.
>
> Jade returns to her seat. "I mean, going around in circles," she replies while extending and moving both of her arms in unison and in a circular motion four times.

Figure 5.1: Jade's hand and arm gesture of rolling in circles.

"What does it mean to slide?" asks Jennifer.

"Not roll!" replies Gage. "It's staying in one place," he says while holding his straight hand above the table. He angles it in a downward position and moves it towards the table.

"Now, you're using your hand and explaining something," says Jennifer looking at Gage. "Do you have a picture in your mind?"

Gage smiles, nodding his head.

"What's the picture in your mind?" asks Jennifer.

Gage replies, "A toboggan[2] going down a hill."

"A toboggan going down a hill!" repeats Jennifer. "Does anyone else have any other pictures in your mind when you hear the word, 'slide,' when you think about what it means to slide?" Nine children put up their hands.

"It's like you're going down a slide," replies Chantal, while holding her left arm up, hand in a downward "pointing" position, and moving it in a swooping motion that ends in a path parallel to the floor.

"Whoosh!" exclaims Kira, as she makes a similar movement with her hand.

Figure 5.2: Kira's hand gesture of sliding down a playground slide.

"And when you go down a slide, what happens?" Jennifer asks, looking at Chantal.

"You stop," Chantal replies.

"You stop at the bottom," repeats Jennifer in a louder voice for the class.

Chantal continues, "And you go straight."

"You go straight," echoes Jennifer while using the students' gestures for sliding. And then, using the students' gestures of rolling, she asks the class, "Are you like a ball that goes over and over and over?"

"No!" giggles the class.

Jennifer continues, "And when you are on a toboggan, you … " She pauses and makes a sliding gesture before continuing: "unless you fall off the toboggan." Using Gage's, Chantal's, and Kira's hand gestures one after another, she says, "So when you slide, you stay upright and you go down."

"What does it mean to 'stack'?" asks Jennifer.

Brittney responds by raising her arms, moving them with a lifting action. "Like lifting up something … chairs."

"Lifting up something," repeats Jennifer, copying Brittney's gestures so that the rest of the class can see. "What do you mean by lifting up something?"

"Like … chairs," replies Brittney.

Jennifer then asks the class, "If Mrs. Wilson asks you to stack your chairs, what does this mean? Can anyone show me?"

Jody holds his arms out straight and parallel to each other, raising them up and down, "Like putting a chair on top of a chair."

"Can you stack these chairs?" asks Jennifer, pointing to two chairs at the front of the classroom.

Jody lifts one chair up and then down on top of the other chair.

Figure 5.3: Jody stacks the chair.

"You're putting a chair on top of another chair … the idea of putting something on something else," says Jennifer as she re-enacts the action of holding a chair and then lifting it on to another chair.

Evolutions in understanding cognition

What can we make of the above vignette? Is it but a charming glimpse into the animated and whimsical world of childhood thoughts and actions – largely naïve, unsophisticated, and inconsequential – when it comes to rolling, sliding, and stacking? Or is there something more profound at play here? In this section we trace a historical lineage of thought about cognition, using it to explore what might be going on in the vignette in terms of thinking, knowing, and doing. This section then serves as backdrop to a subsequent major section meant to situate cognition in a present moment of understandings about bodies that learn.

Cartesian dualism and disembodied cognition

It was in the mid-1600s that Descartes detailed how the then-new philosophy of Rationalism might answer an ages-old question. How can humans be certain of what they know?

Variations of this question are present in the most ancient records of Western philosophy, and many responses have been framed through Plato's "Theory of Forms." That theory asserts that the highest and most fundamental kind of reality is to be found in *ideas* – that is, in an *ideal* realm that comprises non-material, abstract, and unchanging forms. Unfortunately, according to the Theory of Forms, human access to the realm of the ideal must be mediated by physical sensations in the material, changing world. Impressions of the Ideal, then, are necessarily partial, tainted, skewed – imperfect impressions of perfect forms.

With the Theory of Forms as uninterrogated backdrop to his thinking, Descartes argued that the way to certain knowledge was through the renunciation of the flesh (Bernstein, 1983, pp. 115–118). In his second Meditation he penned the agenda for a Western transition into an "Age of Reason" – where reason was contrasted with poorly founded "tradition," "authority," and "superstition" (p. 117). Inspired by Archimedes' assertion that any object can be leveraged if one has access to a single immovable pivot point. Descartes sought an "indubitable" certainty to anchor all of thought. He found this anchor in his *cogito*: "I think, therefore I am." The essence of certain knowing, that is, is thought. Thus, to raise a kingdom of absolute goodness and unquestionable truth, here on earth, humans would have to transcend the body, bracket irrationality, and focus an unwavering gaze on abstract essences (Bernstein, 1983, pp. 115–117). The prize of certainty would be attained in the taming and renunciation of all things tumultuous, unruly, and primitive – that is, all things bodily. Descartes wrote:

> Now, the first and chief prerequisite for the knowledge of the immortality of the soul is our being able to form the clearest possible conception (*conceptus* – concept) of the soul itself, and such as shall be absolutely distinct from all our notions of body... (Descartes, 1641, in Manley & Taylor, 1996, para. 2)

In this way, Descartes effectively locked the dualistic mind–body split into Western thought. Mind reigned supreme in the ensuing Age of Reason and the subsequent Modern Era. On the domain of formal philosophy, Descartes' contribution helped to consolidate and formalize the already-entrenched separation of Ideal (of the mind) and Material (of the body); on the level of popular culture, his disembodied realism became a "common sense" that has since come to underwrite most of educational practice.

Consider, for example, today's prevailing metaphors of learning. Humans "build," "scaffold," "absorb," "give," "share," "acquire," "process," "transfer," "store," and "access" this intangible something called "knowledge." Knowing continues to be imagined as a quality of mind that somehow transcends the bodies that are seen to contain knowledge. In this frame, the body (viz., brain) houses – but does not entail – mind, knowledge, and knowing. Instead, the body is merely the conduit through which knowledge passes and the corrupting and corrupted vessel/container within which it accumulates. Important to this Cartesian frame, knowledge does not begin anywhere in bodies. It originates from and circles back to higher out-of-body places. The highest human quest is therefore to seek knowledge's revelation through empirical inquiry into the world's secrets (i.e., do "science") and then to refine that knowledge by attaining to body-transcending, body-bracketing rationality (do "mathematics"). It is thus that the schools of the Modern Era had no place for the body, as sense-making being with or in the world, in either the bodily training of children in elementary common schools or the spirit-laden mathematics of secondary education.

Returning to our vignette, how might the Cartesian perspective – prevailing well into the 20th century – understand children's self-identification with things that roll, slide, or stack? Surely, to make sense through analogy to experiences of the body could not be to make any proper sense at all. The vignette appears to describe child's play and not the work of education. Such a lesson would have nothing of value to connect to the reasoned sophistication and abstract thought of say, future, transformational geometry – these being topics of mind and not body. In children's elementary education, rolling, sliding, and stacking could only matter in terms of physical capacities that one might want to develop for the repetitive menial tasks of manual work or perhaps of physical fitness.

Cognitivism and its information-processing models

The assumption of a mind–body split reached its apex in the first half of the 20th century with the rise of Behaviorism. In a somewhat ironical turn, given the primacy of the mind in Platonic and Cartesian thought, Behaviorists argued that scientific research is compelled to ignore the unobservable and unmeasurable "black box" of the mind. With the rise of computer technologies in the 1960s, a new science for thinking about thinking became possible: Cognitivism. Capitalizing on a brain-as-computer metaphor, Cognitivism promised to restore mind (i.e., as software in the brain's hardware) to the picture after a half-century of exile in scientific research.

How would a Cognitivist narrate the events of the vignette?

A Cognitivist read would likely be structured around the flow of information, attending to inputs, outputs, channels of communication, processing, storage, and a cascade of other entailments of a computer-based metaphor. In this frame, the teacher might be seen as a programmer, and the classroom conversation might be likened to the flow of thoughts as information or data. The teacher's prompts would be to help the children retrieve long-term memories of prior experiences of rolling, sliding, and stacking. These memories could then be processed into revised meanings by "re-presentation" and "abstraction" and returned to brain storage for future use. Physical gestures would focus children's attention and improve the virtuosity of input. Engagement of motor as well as executive brain regions, along with more time on task, would ensure adequate processing time, which would in turn increase the probability of memories moving from short-term to long-term memory. The careful use of vocabulary, these attached to motor cues, would enable learners to multiply tag their memories for efficient and reliable retrieval later. In brief, the thinking about thinking shifts toward sweeping inferences about the workings of the still-unobservable-black-box as if it functioned like a computer.

In restoring the topic of mind to research into human knowing, then, the Cognitivist information-processing model actually introduced an even more insidious way to skirt the issue of the body. Knowledge, conceived through a computer metaphor, continued to be understood as something accumulated, where the good teacher is one who facilitates a stimulating environment to make knowledge endure and thus become retrievable from memory.

This computer metaphor is alive and well today. In true Cartesian fashion, knowledge continues to be positioned out in the world awaiting proper transfer into learners' minds/brains. This belief conditions the very teaching acts one endorses. From such a perspective, spatial acts of self in and with world (including first person experiences of oneself, of extensions of self, and of acts of self with and upon things and others – these in mutually informing conversation with second person viewings of self, others, and things in world) can only be peripheral and incidental to the conceptual understanding of the mathematical transformations of rolling, sliding, and stacking. Moreover, in this view, to attend directly to such body-in-with-space engagements would simply constitute an unnecessary and unwieldy addition to mathematics curricula.

The rise of cognitive science

The brain-as-computer metaphor of the mid-20th century, while still a prominent and popular notion in the western world, has been subject to widespread negative critique over the past several decades. For the most part, these critiques have arisen out of the computing sciences – more specifically, research into artificial intelligence. The confident predictions of the 1950s, that computers would soon dwarf the flesh-based intelligence of humans, proved to be more the stuff of fiction than of science. Researchers in human cognition were forced to rethink their assumptions about the human mind and body. How could it be that the things that highly educated, disciplined minds found difficult (e.g., performing calculations, solving differential equations, playing chess) could be so readily performed by electronic brains, yet so many competencies mastered by young children (e.g., recognizing faces, understanding language, manipulating others) were well beyond the capacities of the most sophisticated machines?

The answer lay in the slow realization that the self-modifying human brain is not a piece of hardware and knowledge is not software or information. More concisely, the assumption of a mind–body separation is untenable, and this assertion is supported by a half-century of neuroscientific research. Still, the assumption of a separation continues to thrive in the cultural stock of knowing, maintaining a vibrant currency in the discourses that surround teaching and learning.

Cognitive science offers an explanation for the persistence of this troublesome assumption. It is difficult to interrupt the notion of disembodiment when virtually every reference to knowledge and learning appeals to images of separation and transfer. A constant barrage of expressions consolidates the dominant metaphors of "knowledge as a stable, constructible object" and "learning as acquisition," and these habits of interpretation actually affect the structure and activity of the brain. Cutting the story much too short, interpretative habits become neural proxies for the physical actions that generated them and for which they stand (Lakoff & Johnson, 1999; Varela, Thompson, & Rosch, 1991).

This sort of realization has helped to trigger growing, yet varied, empirical and theoretical research that rejects Descartes' duality and seeks to understand the mind as embodied and the body as minded. We now turn to that literature.

Embodiment

The embodiment literature, diverse though it has become (Gerofsky, 2014), could be said to have roots that reach back to the 1700s to the philosopher,

Giambattista Vico. As a formal domain, it dates at least to the middle of the 20th century, to the philosopher, Maurice Merleau-Ponty (1945, 1962) figuring prominently. Merleau-Ponty emphasized the primacy of perception and, in so doing, identified the body as the site and means of knowing. This ontology of "the flesh of the world" (*la chair du monde*) (Merleau-Ponty, 1964) stood in stark contrast with Descartes' philosophy, which posited human perception as a carnal, readily deceived, error-introducing *channel* from the outside to the inside. It was not to be trusted. For Merleau-Ponty, sensory perception was not a channel that separated, but a membrane that connected. Perception, he argued, is an integral part of cognition. More broadly, and in an explicit rejection of other prominent dichotomies such as self–other and knowing–doing, Merleau-Ponty positioned perception as a key element of action-in-the-world. For him, action was not a reflection of knowing; it was an aspect of knowing. The body was not a vessel for knowledge; it was a knowing system.

Later Maturana and Varela (1972, 1991) put forth the idea that our actions created in the flow of knowing become our ways of being – "the realization of living," as noted in the subtitle of their first book. Varela coined the term, "enactive" (Maturana & Varela, 1991, p. 255) to communicate the view that the very roots that give rise to perception, movements, and object manipulation also enable and specify conceptualization and reasoning: "Cognition depends upon the experiences that come from having a body with various sensorimotor capacities," and "perception and action are fundamentally inseparable in lived cognition" (Varela et al., 1991, p. 173).

Varela's Enactivism is just one of many theories within the embodiment literature. Over the past decades, more recent theories of embodiment have arisen in other domains such as psychology and sociology. Many of these have been taken up and elaborated in mathematics education research – although, as Gerofsky (2014) noted, interpretations and emphases have varied dramatically within the field. Despite disconnects and occasional discontinuities, all of these theories converge on one critical point: the rejection of a mind–body separation. As well, several of the theories extend the discussion beyond the biological body and address interactions among other bodies and the environment, including accounts of language and other cultural phenomena.

Taking seriously the idea of embodiment – that is, that humans necessarily exist all-at-once as dynamic bodies within other clusters or networks of bodies on micro and macro scales such as bodily, biological, social, cultural, and evolutionary – provokes additional questions, such as: In what ways does an embodied cognitive perspective necessitate a complex understanding for the body *in* mathematics as well as the body *of* mathematics (de Freitas & Sinclair, 2013)? What critical insights do these ways of thinking offer us in making sense of young children's spatial reasoning in mathematics? And how does this compel a turn to spatial reasoning in curriculum? These questions comprise the focus of the remainder of the chapter.

Bodies *in* and *of* mathematics

For the past 20 years in the field of mathematics education research, there has been considerable interest in recognizing and understanding the diverse ways that the body is central to and subsumed in learners' mathematical cognitive development (e.g., Núñez, Edwards, & Matos, 1999; Nemirovsky et al., 2004; Radford, Edwards, & Arzarello, 2009; Campbell, 2010; Stevens, 2012; Edwards et al., 2013). With this emphasis also comes the need to attend to the dynamical growth of mathematics concepts in the study of students' learning (e.g., de Freitas & Sinclair, 2013; Thom, 2012; Towers & Martin, 2014).

The emergence of a theory of embodiment turns out to be much more than a theory of how individuals' bodily action informs their emerging understandings. Consider current fields of study within the embodiment literature: In complete rejection of Cartesian duality, these researchers tend to insist that mathematical understandings rooted in the biological body cannot be considered apart from the epistemic body of mathematical knowledge in which those understandings arose. We see an insistence on the simultaneous consideration of knowing bodies within bodies of knowledge and this applies to bodies at many levels of systemic organization, including individual learners, groups of learners, and cultures. Moving far from the common notions that mathematics is ideal – that is, that objective knowledge exists independent of knowers and that learning is a process of acquiring or internalizing that knowledge – the thinking here is that engaging with and in mathematics affects and is affected by the other bodies and any perceived separations are merely heuristic conveniences. Consider the example in which Jade develops of the concept of rolling. The concept of rolling emerges as she moves her arm in a circular motion and then identifies the sphere as a ball and as an object that rolls. Not only does she expand this concept for herself, at the same time, she does this for the larger body of learners of which she is a member. Here, the idea of rolling takes shape, at least in part, with the action she performs with one arm, pointing to the sphere, and finally through the actions performed with both arms. This constellation of actions occasions particular spatial ideas that can be understood as relational responses situated across the system that constitutes Jade, the sphere, the teacher, the other students, the classroom, and the concept of rolling.

Furthermore, following a point made by de Freitas and Sinclair (2013), when mathematical technologies such as tools, symbols, and diagrams are not taken for granted as inert artifacts of the external world, but are reconceptualized as animate bodies integral to larger networks of humans and mathematics, such technologies can be conceived as having the potential to be dynamic entities (cf., Pickering, 1995; see also Chapter 6, which provides some examples of the use of digital technology for spatial reasoning in the early years). When the use of technologies provokes re-mapping in the brain, such objects co-emerge and co-evolve with our body or bodies in what Malafouris (2013) describes as "a continuous and interactive coordination between neural and extra-neural physical resources."

To re-emphasize, embodiment moves far beyond a theory of how individual bodily actions inform emerging understandings. It is a theory of radical embeddedenss – of how there are no sharp boundaries to bodies, of how bodies are nested in bodies, of how bodies of knowledge unfold from and are enfolded in the bodies of knowers, and of how knowing, doing, and being are ultimately indistinguishable aspects of the same whole.

Bodies *in* mathematics

Given our historical lineage through duality and a Western privileging of the individual as the agent of sense making (Bowers, 2011; Henrich, Heine, & Norenzayan, 2010), it seems obvious – even logical – to identify the body or bodies in the vignette as independent, non-intersecting agents (and Jennifer as their similarly discrete teacher). Even so, by interpreting the vignette from different perspectives within embodied cognition, we are able to recognize and articulate specific spatial manners in which the bodies of the students (and Jennifer) work in the geometry lesson.

Re-cognizing[3] each human body, in mathematics, as a dynamic cognitive system

From an enactivist perspective, the body is conceived as a dynamic cognitive system in which knowing is understood as effective action and thus, it is not possible to separate one's or others' knowing from doing and vice versa. In the words of Maturana and Varela, "all doing is knowing and all knowing is doing" (1991, p. 26). Looking at the lesson from this perspective, we see the children and Jennifer use their bodies to move, vocalize, and gesture about how objects roll, stack, and slide. The children and Jennifer's physical *actions* serve as observable and communicative *knowings* (Pirie & Thom, 2001; Thom, 2012) of which and by which the children's spatial reasoning grows. It is the work in which their bodies engage that enable the students and Jennifer to see, speak, gesture, move, visualize, and imagine the geometric transformations of the 3D objects. These are the very ways in which conceptions of sliding – such as not rolling, going straight, not turning over and over, staying upright, and moving or swooping downward – are worked out, articulated, and justified (Radford, 2014; Radford et al., 2003). Conceptualizing the body, in mathematics, as a dynamic cognitive system enables students and teachers' physical, visual, verbal, written, mental, and (in)formal activity to be taken not simply as representations of abstract spatial concepts but, importantly, as corporeal and contextually grounded forms of cognition (Pirie & Kieren, 1994; Davis, 1996; Davis, Sumara & Luce-Kapler, 2008; Radford, 2014).

Consider another example that illustrates these points in more detail. Thom and Roth (2011) reported on how a child's body worked to spatialize the faces, edges, and vertices of two different rectangular prisms:

a   b

Figure 5.4: Photographs of the two rectangular prisms.[4]

Upon first glance, it seems as though the child, Owen, looks at and touches the block's faces, edges, and vertices as he counts them one by one. However, taking a much closer frame-by-frame examination, what appears to be Owen's hands fumbling to achieve one-to-one correspondence with the block's attributes is in fact the orienting and re-orienting of finger(s) and an entire hand that generates three distinct hand positions – a *flat* hand on the block's 2D faces (Figure 5.5a–c), a *straight* finger on the block's 1D edges (Figure 5.6a–i), and a finger*tip* on the block's 0D vertices (Figure 5.7a–b) – with which these attributes are then counted. Importantly, in addition to hearing Owen's numerical knowing and to observing his counting actions, we note his attention and exploration of the blocks – their faces, edges, and vertices – as signaling a rudimentary experiential basis in connecting visual and kinesthetic conceptions of spatiality.

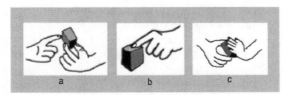

Figure 5.5: Owen's different hand positions as he counts the wide block's faces.

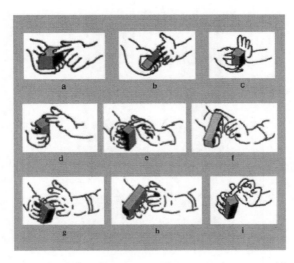

Figure 5.6: Owen's nine hand positions as he counts the narrow block's edges.

Figure 5.7: Owen's two hand positions as he counts the narrow block's vertices.

It is important for the reader to bear in mind that objects such as the rectangular prism do not inherently exist as the block that Owen holds, or as words on this page, or as the 2D diagrams and photographs (e.g., Figure 5.5) that appear in this very book or even in school textbooks that represent the rectangular prism as a 3D object in 2D. Rather, the rectangular prism as a mathematical object transcends the block, the words, the diagrams, and the photographs. The rectangular prism emerges as the result of the particular and collective distinctions (e.g., a regular hexahedron with six identical square faces) that we make based on our biological and culturally embedded experiences of bringing forth a world of [mathematical] *significance* with others (Maturana & Varela, 1991).

In a different lesson, a year later, Owen explores how he can produce 2D shapes using a wooden cube and the overhead projector (see Figure 5.8). Three specific moments occur as Owen looks at the screen and watches for the 2D shapes that appear as he places the cube on one of its faces (Figure 5.8a-b), holds it on one of its edges (Figure 5.8c–d), and then positions it on one of its vertices (Figure 5.8e–f). Here, we see transformations taking place not only with the spatial–geometric emergence of particular mathematical ideas, but with the cube itself as Owen works with the block in ways that simultaneously real/ize it as a 3D object and 2D square, oblong, and hexagon.

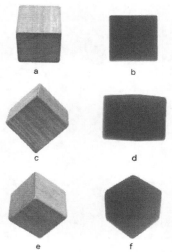

Figure 5.8: Photographs of the three different positions that Owen holds the block on the overhead projector and the resulting 2D shapes that appear on the screen.

In relation to these two episodes, it is possible to understand touch and movement as mutually entwined processes. Each informs the other.

Further still, in the opening vignette of the geometry lesson, the children and Jennifer's bodies are rich with touch and movement that involve physical and re-membered objects such as a sphere, a ball, a slide on the playground, a toboggan, and chairs. Exactly what the body encounters when it meets the object can be conceived as the object's response in relation to the body's action of touching and moving. Gage articulates this well when he explains that when you are tobogganing, your body, which is in contact with the toboggan, is "staying in one place" – that is, on the toboggan, and it is the toboggan that is actually sliding down the hill. These understandings demonstrated by Owen and the class of 2nd-grade children cannot be taken as fixed abstractions of spatial concepts or even conceptions that are inherent in the objects themselves. This is because these kinds of bodily experiences and knowings as dynamic sensorial whole-body happenings – of sights, scents, emotions, feelings, sounds, vibrations, movement, velocity, angle, pressure, acceleration, momentum, friction, and so on – exist all-at-once and in ways that depend on their bodies to touch and move with the world as much as the world does with their bodies. "Indeed, there would be no space, no concept of space or of being 'in space,' no objective 'out there' short of movement to begin with" (Sheets-Johnstone, 2012, p. 396).

Re-cognizing each individual human body, in mathematics, as an expressive body

From an embodied cognitive perspective, bodily experiences such as sliding, rolling, and stacking, are considered fundamental for the development of concepts. Fauconnier and Turner (2002) used the term "conceptual blending" *to explain* conceptual development. Conceptual blending plays a critical role in human reasoning and actions. It is purportedly responsible for the origins of human achievements such as language, science, religion, and arts. Lakoff and Núñez (2000) described conceptual blending as "the conceptual combination of two distinct cognitive structures with fixed correspondence between them" (p. 48). Conceptual blending allows for both the conceptualization of the original embodied experiences and the creation of new concepts blended from other concepts. For instance, turning to the vignette and the discussion about sliding, Gage makes reference to "a toboggan going down a hill." Here, the re-membered experience of sliding down a hill on a toboggan – palpable and personally experienced – is blended with the geometric transformations of mathematics. This blending can thus contribute to the development of concepts of sliding 2D shapes and translating 3D objects.

Additionally, the vignette provides examples of how gestures that include body movements and voice are used to talk about sliding, rolling, and stacking. For instance, Kira exclaims "Whoosh!" and accompanies this with a hand gesture as an immediate response to Chantal's statement and action: "It's like you're going down a slide," whilst her hand swoops downward. The body becomes a means to communicate memories that blend with other 3D objects. The body, thus engaged, gives rise to actions that, thought together with geometric concepts, contribute spatial meaning to those concepts.

Re-cognizing the bodies beyond the individual learner in mathematics

Embodied cognition in a radical sense is the idea that knowing and knowledge is distributed within and across systems. So while we may zoom in and focus on understanding the spatial reasonings of individual learners, it is the very idea of cognition as distributed across larger bodies that compels us to ask how spatial reasoning takes place within and on different levels where mathematical bodies extend and comprise beings beyond edges limited by the skin of the individual learner.

Drawing on Merleau-Ponty's (1962) example of a blind man with a white cane and similar to Bateson's (1972) argument about the arbitrariness of imposed boundaries, de Frietas and Sinclair (2013) explain that nonhuman objects and mathematical concepts can be understood as extending the body beyond individual learners. For example, when Jody lifts the chair up and puts it on top of the other chair, it is not simply Jody's body that is involved in the spatializing of mathematics or even that Jody spatially acts upon the environment. Rather, it is with the chairs that he acts and, in so doing, it is the concept of stacking that he articulates and makes available for himself and the class. Effectively, we may identify the body in this mathematics as Jody–chairs–stacking. And it is possible for us to inquire into how these different bodies work as a larger complex whole as well as how the systemic manners and the kinds of spatial reasoning emerge as Jody stacks the chairs. Indeed, when we look at the entire vignette, it is difficult if not impossible to separate the concepts and conceptions of rolling, stacking, or sliding from: the children's and Jennifer's bodies; the socio-cultural practices in which they are situated; and the contexts of tobogganing, rolling a ball, going down a slide, or stacking chairs – much less the objects with which these spatial transformations are brought into being.

Re-cognizing bodies, in mathematics, as a coherent collective

In a different way, the idea of cognition as being distributed across agents without the exclusion of the environment allows us to move beyond the level of the individual learners to socio-cultural levels where we can re-cognize groups of learners in mathematics as itself, a coherent collective agent. For example, it is possible to conceive Jennifer and the students as a *class* engaged in mathematics. On the socio-cultural scale, the body in mathematics is neither each individual learner nor Jennifer as their teacher or even the class as a teacher with a group of students. Rather, the class considered on a different scale, and over moments in time, might act and be understood as a cohesive body with particular ways of working in the context of the geometry lesson. Therefore, in re-cognizing the class as a potential collective body engaged in mathematics, we can ask questions about the class, and how it works as a body in and of itself during the lesson with respect to spatial reasoning.

It is at the level of the collective *qua* body that we are able to focus on particular dynamics. In the vignette, we see definite moments in which the class "co-acts" (Martin & Towers, 2011; Martin, Towers, & Pirie, 2006) in coordinated and responsive ways such as when the class articulates sliding as a toboggan going down a hill, going down a slide, and "Whoosh!" In contrast, there are other instances where certain events happen and the class does not proceed with their work but rather, changes tack. For example, what precipitated the class conversation about rolling, stacking, and sliding was the fact that the class could not continue with the task of predicting what the collection of objects could do without shared (i.e., distributed) meanings of the concepts of roll, slide, and stack with which to reason. In a similar manner at the level of the individual learners where the growth of their spatial reasoning is dependent on and contingent upon previous knowings, it is the class' collective spatial reasoning that depends on and is contingent upon the members' actions in the group (Towers & Martin, 2014).

Inter/intra-body conceptualizations

The juxtaposition of spatial concepts

The very idea of juxtaposition implies the effect of contrast that happens when two or more seemingly disparate things are brought together. Nemirovsky and Ferrara (2009) described the juxtaposition of mathematical concepts as "juxtaposing displacements" or "cubist composition." Characterized as the enaction of partial aspects of context(s), regardless of time and location, and occurring adjacent to one another, we see the class as juxtaposing ideas of sliding and what it means to slide and creating a coherent enough concept that effectively spans different contexts, time, and space. In the geometry lesson, as the students and Jennifer explore what it means to slide, the class brings together conceptions of spatial transformations of going straight, not rolling, a downward movement, not like a ball that goes over and over, tobogganing down a hill, going down a slide, staying upright, and a swooping motion. Importantly, we also see the children's concepts as unpredictably dynamic and ever-evolving in the immediacy of their reasoning. This is what keeps these concepts, and the ways the children work with them, open to change.

The collectivity of spatial concepts

The idea of collectivity within enactive and complexivist perspectives may also be used to interpret the spatial concepts explored by the class in the vignette. Here, personal and collective conceptions are assumed to be interrelated, culturally embedded, and integral to mathematical concepts. Moreover, collective understandings are those that happen at the level of the group or class. These occur at times when an idea or action is taken up by other members of the group, is revised, expanded on, and becomes vital to the group's integrated way of thinking about the mathematics. In the vignette, the idea of tobogganing is picked up by the class, compared with rolling, integrated with other conceptions of sliding, incorporated eventually as a concept by the class, and then, later used to predict the potential ways in which 3D objects move.

Moreover, we see how the concepts of rolling, sliding, and stacking play out in emergent and unpredictable ways as they are blended in relation to the spatial transformations of 3D objects and combinations of spatial transformations. Effectively, the three concepts that involve movement which are associated by the children with a ball, toboggan, slide, and chairs are brought into being by the children as they play with these spatial ideas, bump them against one another, and through a process of either cohering or discarding, the ideas of rolling, stacking, and sliding undergo changes that eventuate the mutual transformation of the class and of the mathematics. We take these group dynamics as playing an integral role in how the class makes spatial sense of the three concepts. It is also clear that occasioning robust collective spatial reasonings with young children is not just a matter of making concepts and their meanings available to the class or by amassing a collection of student conceptions. Rather and in a profound way, the collectivity of spatial concepts speaks to the coherence that eventuates from the spontaneous, moment-to-moment collaboration of the group's thinking as a complex integrated whole.

Bodies *of* mathematics

Re-cognizing the body of mathematics as a body of knowledge

If we consider mathematics as a human activity, as opposed to a collection of actual objects independent of our being human, then mathematics is a body of knowledge entwined with culture, including language. In particular, the use of metaphors informs our understanding of mathematical concepts. As already mentioned, the process that makes it possible to conceptualize from metaphors is called "conceptual blending": the combination of distinct cognitive structures that give rise to the creation of new concepts. The extensive work on metaphor, as cognitive process, is rooted in neuroscience. Neurolinguist Pulvermüller explained that "language is 'woven into action' at the level of the brain" (2011, p. 423). Whereas the link between words and the objects to which they refer are considered "a result of associative learning driven by correlated neuronal activity" (p. 423), the understanding of actions words – their referent actions having occurred in interactional contexts – calls up the action systems used to generate the actions in the first place. There are many examples of this phenomenon in the vignette. For example, upon hearing the word "stack," Brittney immediately comments, and thus the motor and premotor areas associated with stacking are activated in the brain (Pulvermüller, 2011). In other words, we make meaning through experiences in perception and action: we make sense by engaging the senses. The important point of language is that words – though thought about in terms of symbolic representations – acquire currency and meaning not in the symbolic but rather in the sensual experiences of perception and action to which they have been neurologically linked in the first place and that subsequently come to be called up and iteratively modified in each subsequent invocation.

Metaphors are much more than linguistic expressions. They reflect underlying conceptual mappings (physiologically entailed because of neurological precedence) and in a profound manner, according to Lakoff and Johnson (1999) as well as Fonagy and Target (2007), they become the "architecture" of mental life. If we understand reasoning as sense-making that goes back to and depends upon prior experiential sensing, then in the particularity and seemingly abstract forms of mathematics, metaphorical language becomes vital to such bodily experience (bodies writ large) and subsequent conceptual understanding. It is through the body that a spatial world comes to be sensed, known, realized, and reasoned about and through. As such, strong spatial reasoning, arguably, is seminal to strong mathematical knowing and should figure central in any mathematics curriculum.

Chapter 1 provided examples of the use of metaphors in mathematics. Simply put, these authors assert that metaphors provide the conceptual basis of mathematics. Consider, for example, the words used for logic and set theory (Lakoff & Núñez, 2000). These words relate spatially to the *container schema* and to topological associations: interior, exterior, and boundary. Interior and exterior are words embedded in the language for set theory and Boolean algebra, and this enables symbolic logic in mathematics. A very specific example in set theory is the expression "the element x is in set A."

Another important aspect relating to mathematics as a body of knowledge – mentioned in Chapter 1 and recurrent throughout the book – is the embodiment of conventions in pictures and diagrams. These images, combined with their linguistic associations, are used to define and provide cultural meaning in mathematics. As conventions, the meanings are not inherent in the pictures and diagrams themselves but rather, are learned and carried forth from one generation to the next through participation in the practice of mathematics. Chapter 8 provides examples of cultural conventions and norms used to represent 3D objects through 2D images.

With respect to the authors of Chapters 1 and 8 and those authors, throughout the book, who write of young children working in 2D and 3D contexts, we wish to stress the following: how we come to know what we know is so embodied in our very ways of doing and being with the world that we take those bodily roles for granted – even to the point of seeming inaccessibility to our awareness. For example, when we encounter a rectangular prism – either by looking at a static 2D diagram, a photograph (e.g., Figure 5.4), the printed words "rectangular prism," or even a physically present, wooden block – it is easy to lose appreciation for the fact that our understanding of such representations necessarily implicates our body in radically profound ways.

At a finer grained detail, consider our ability to look at a photograph of a block and be able to identify it as a rectangular prism. Doing so requires the unconscious movement and experience of our eyes, of knowing what to do. Such movement and experience not only invokes a recognition of straight, parallel, and perpendicular lines but the recognition of certain shapes and the comprehension of the three-dimensionality of the block in the photograph – that comprehension itself signaling a necessary spatial coordination of 2D and 3D. Further, recognizing the 2D photograph of the 3D object as a 3D rectangular prism requires specific cultural know-how of "that which is necessary for us to ignore" and "that to which we must pay attention." For example, we know to ignore the fact that the lines appear on a 2D page and instead, "read" or interpret the lines of the photographed object as if these were in a 3D space so as to "see" the lines as edges that define faces, visible and not visible in the photograph, as being rectangular and either parallel or perpendicular to five other faces. In this way, we interpret the image as if it were indeed a 3D rectangular prism. The ability to look and see in spatially specific, mathematical ways is not just embodied in eyes that unconsciously know what to do (Merleau-Ponty, 1962). In a mutual way, it lives in cultural forms of mathematical knowing that tell us how to interpret the photograph (Vygotsky & Luria, 1993).

Consequently, the selection of metaphors, images, and even gestures used to provide meaning to mathematical concepts poses important pedagogical implications. An example of this is the metaphor of number as a collection of objects (Lakoff & Núñez, 2000). Although this is a common way to introduce the meaning of number in schools, the metaphor does not lend itself easily to the concept of fractions and negative numbers. The number line, in contrast, supports the understanding of numbers as location, direction, and movement, as well as operationalizing them – such as in the multiplication of a negative number. In fact, the research studies described in Chapters 3 and 6 suggest that the development of earlier intrinsic-dynamic spatial skills has a positive impact on the performance of learners' mathematical skills.

The body of mathematics proper and the indeterminate potentiality of mathematical concepts

In the classroom, mathematics is commonly regarded, and thus treated, as static content. Consequently, mathematics is relegated to a passive role in children's learning (de Frietas & Sinclair, 2013). It is not itself a growing, changing body (Jardine, 1994; Thom, 2012). This concern becomes particularly poignant when we think of early years school mathematics that, as elaborated in Chapter 2, tends to be substantially limited to static skills. We find this striking given the research presented in Chapter 3 that shows a correlation between the development of dynamic spatial reasoning in the early years (i.e., mental rotation) and ensuing performance in different aspects of mathematics (i.e., arithmetic computation and problem solving).

In our final paragraphs, we turn to the notion of "indeterminate potentiality" to further extend the conversation in light of the potentiality of the body of mathematics at the interplay of young children's imaginings and understandings of spatial concepts.

Whereas some theories of embodied cognition (e.g., systemic ecology and enactivism) conceive the mathematics classroom as an ecosystem, doing so does not dismiss the structure and organization inherent in these spaces. Importantly, structure and organization can be conceived and enacted in ways that are dynamic and emergent. Conversely, while the idea of an ecosystem shapes structure and organization as fluid and open to change, it also calls up these notions in terms of stability. The same can be said about school mathematics.

As an animate, organic body, school mathematics – and more specifically spatial concepts within a mathematics curriculum – can maintain particular structural relationships while still being flexible enough to also allow those relationships to be conceptualized as multiple and indeterminate (Châtelet, 2000; Sinclair, de Freitas, & Ferrara, 2013). For example, in Chapter 7 we encounter instances that elucidate the ways in which drawing – as both emergent acts and emerging artifacts – develops and communicates mathematical actions in spatially dynamic, multiple, and indeterminate ways.

In the vignette, the very reason for the class' conversation is to explore the indeterminate potentiality of what it might mean to roll, slide, and stack. The students are invited into, and engage in, an animated discourse about rolling, sliding, and stacking. They are invited to do so for the purpose of thinking more deeply about these notions and to, if necessary, bring into question their taken-for-granted assumptions. In this manner these bodies – individually, collectively, and together with other bodies implicated in and by these spaces – flesh out the possible ongoing and recursive reconceptualizations of rolling, sliding, and stacking. These kinds of opportunities naturally lend themselves to exploring the potentialities of concepts in general – concepts at play for instance in such questions as "Is a half of a half *really* a half?" (Thom, 2012, pp. 347–359). Put differently, and in the context of mathematics philosopher, Gilles Châtelet's (cf., de Freitas & Sinclair, 2013) image of the number line as being elastic (and therefore indeterminate in length), such activities make it possible for students to "carve out the virtual real numbers embedded between whole numbers by grabbing and stretching the number line so that it brings forth an infinitude of numbers that were imperceptible a moment earlier" (pp. 465–466).

Linking ideas

Chapters 4 and 5 represent efforts to address the question, "If spatial reasoning is so important, why has it taken so long to be noticed?" The response offered in Chapter 4 focused on cultural conditions that compelled aattention to utilitarian knowledge, measurable achievement, and standardized outcomes. In this chapter, we addressed the question by delving into deeply entrenched beliefs about knowledge and learning. We traced several familiar but varied interpretive lenses on a teaching and learning event. In the doing, we drew attention to the continued presence of disembodied Cartesian dualism in contemporary curricula where mathematics lives on, uninterrupted, as an objective and objectified assemblage of knowledge.

At the same time, we wish to emphasize that the very phrase "spatial reasoning" suggests a bodily engagement in space that places importance on the senses: While the sense of sight and the manipulation of objects, both physically and mentally, are a main focus in Chapters 2 and 3, we have, in this chapter, considered the perceptions and actions of the bodies in and of mathematics in order to extend the notion of embodiment beyond the biological body. At the risk of overstating the point, we would invite readers to consider using the phrases "spatial reasoning" and "embodied mathematics" interchangeably as they read other portions of this book. Importantly, we do not mean to suggest that the phrases are synonymous; on the contrary, and as developed in other chapters, there are significant differences. At the same time, there is significant overlap. A child who engages in the act of mentally rotating a shape is not just performing an act of spatial reasoning. She is demonstrating embodied mathematics. She is recursively enacting her embedded and embodied knowing. Spatial knowing, doing, and being as radically embodied is about how our bodies engage with the world, regularly sensing and making sense of the situations in which we find ourselves.

Such is the sensibility that frames the next three chapters where the discussion moves into the space of mathematics pedagogy. In brief then, desires for and efforts toward the development of spatial reasoning in the context of mathematics learning are simultaneously moments of embodied mathematics.

Notes

1   This research was supported, in part, by grants from the Social Sciences and Humanities Research Council of Canada. We thank the teachers, students, W.-M. Roth, J.-F. Maheux, and the assistants who contributed to the research.

2   A toboggan is a simple snow sled that was traditionally used by the Innu and Cree of Northern Canada as a means of transportation. Today, children use toboggans for recreation during the winter season to go down snow covered hills.

3   From the Latin word, *recognoscere*, we use the term, re-cognizing, not in the sense of its usual everyday meaning, which is to "identify as already known" but in its second meaning to "know again" anew.

4   Figures 5.4, 5.5, 5.6, and 5.7 appear on pages 270–274 of Thom and Roth (2011).

Section 3

What are the curricular and pedagogical implications of spatial reasoning?

SECTION COORDINATORS: KRISTA FRANCIS, MICHELLE DREFS

Chapter 6: Spatializing the curriculum

Children are capable of engaging with and understanding sophisticated geometry ideas when we harness and infuse spatial reasoning into the learning situations. In this chapter we look across a series of classroom vignettes that illustrate how thoughtful materials and innovative tasks in geometry enable children to build their understanding in the complex context of the K–2 classroom.

Chapter 7: Motion and markings

This chapter focuses on the dynamic dimension of spatial reasoning, particularly in terms of how drawing is used to develop and communicate mathematical ideas. Drawing is a way in which children explore and become aware of spatial concepts and relations in their dynamic, embodied and 3D world, and also a way for them to record and reflect on that motion.

Chapter 8: Interactions between three dimensions and two dimensions

To effectively use spatial reasoning, children need to learn the conventions and interpretations of the 2D representations of the 3D objects and be able to move fluently between each dimension. 2D representations are commonly uncritically allowed to stand for the 3D object. This chapter explores the implications of this oversight.

6

Spatializing the curriculum

CATHERINE D. BRUCE, NATHALIE SINCLAIR, JOAN MOSS, ZACHARY HAWES, BEVERLY CASWELL

— In brief …

Children are capable of engaging with and understanding sophisticated geometry ideas when we harness and infuse spatial reasoning into the learning situations. In this chapter we look across a series of classroom vignettes that illustrate how thoughtful materials and innovative tasks in geometry enable children to build their understanding in the complex context of the K–2 classroom.

Introduction

Recent large-scale studies have demonstrated that early mathematics is a strong predictor of overall school success (Duncan et al., 2007; Claessens, Duncan, & Engel, 2009; Claessens & Engel, 2011). Smaller-scale studies have identified spatial reasoning as a critical form of mathematical thinking that provides the foundation to young children's mathematics learning (Verdine et al., "Deconstructing Building Blocks," 2014). For example, in their study with very young children, Farmer et al. (2013) found that spatial reasoning at age 3 was a better predictor of general arithmetic abilities at age 5 than language (vocabulary) or even general math skills (see Chapter 3 for more details). Research further suggests that spatial reasoning skills play an important role in predicting which students enjoy, enter, and succeed in STEM disciplines and the arts (Newcombe, 2010; Diefes-Dux, Whittenberg, & McKee, 2013). Given that recent research clearly confirms the malleability of spatial reasoning (see Chapter 3), efforts to increase children's abilities to reason spatially make this area of research particularly important and timely.

Unfortunately, current curricula and overall mathematics instruction rarely focus on spatial reasoning as a curriculum goal, and the range of learning activities for young children that foster spatial reasoning are sorely lacking (see discussion in Chapter 4). Researchers of the Spatial Reasoning Study Group have invested energy in developing classroom-based tasks and interventions, which continue to be field-tested in collaboration with educators. These tasks, lessons, and interventions are based on the laboratory-based research findings presented in Chapter 2, but have been adapted in ways that make them appropriate for the everyday complexities of the classroom. In this chapter, we look at three different classroom vignettes from these classroom-based research projects, and identify what spatial reasoning is being mobilized in each instance. The topics in focus are: symmetry, mental rotation, and congruence with physical transformations. All three examples have commonalities worth noting. First, they oscillate between and among working with concrete materials, the imagination, and considering abstract ideas in playful contexts for young children. A second common theme across all three vignettes featured is an explicit attention on how multitouch technologies and mathematics tools play a large mediating role in supporting: (i) the actions of spatial reasoning such as transforming and symmetrizing, (ii) deeper understanding of spatial ideas, and (iii) more precise discourse amongst the children and with the teacher. The third common theme is that all three examples push the boundaries of typical geometry curricula, moving beyond the usual naming and sorting of shapes. In the vignettes presented, we address atypical expectations, or learning outcomes, in greater depth and with a dynamic typology underpinning task design. We focus on geometry, a common strand of mathematics taught in school, in order to illustrate how to extend, enliven, and enrich the curriculum and children's thinking with spatial reasoning as the foundation.

Exploring the actions of symmetry

Symmetry is a central idea in mathematics; it is an object of study in and of itself, but it is also used as a powerful way of solving problems (Leikin, Berman, & Zaslavsky, 2000; Stewart, 2007). While often treated as a static concept, the notion of symmetry as a transformation requires a more dynamic approach in which a particular motion is required to transform one initial figure into a final one. While everyday experiences are filled with static symmetric shapes (hearts, stars, etc.), they do not, in general, involve recording consecutive moments of an object's movement. Thus, the teaching and learning of symmetry requires tools, tasks and communicative opportunities for children to focus both on the action of symmetry transformations and the result of such transformations.

Symmetry is an area of spatial relations where children have a great deal of embodied knowing long before they learn about them formally in geometry classes. For example, there is evidence that children spontaneously construct symmetrical figures during informal play at the pre-school age (Seo & Ginsburg, 2004). Indeed, in his review of children's understanding of space and its representations in mathematics, Bryant (2008) suggested that, "one of the most important challenges in mathematical education is how best to harness this implicit knowledge in lessons about space" (p. 3). Even when symmetry is studied, the focus has been placed on the difficulties that children face when identifying symmetrical figures and on their confusion between symmetrical (usually around a vertical axis of symmetry) figures (Bryant, 1969, 1973; Boswell, 1976; Bomba, 1984; Quinn, Siqueland, & Bomba, 1985). These studies were concerned with "the development of children's ability to tell symmetrical figures apart, not to understand the relation between them" (Bryant, 2008, p. 34).

In order to approach the teaching and learning of symmetry in a deeper way that would engage spatial reasoning, we designed a three-part lesson that centrally features the dynamic geometry environment *Sketchpad* and that was aimed at Grades 1–3 children. In the following section, we briefly describe the three lessons, and include particular moments in each that highlight the features we find important in the children's thinking.

Highlights from the three lessons

After an initial survey of the children's understanding of the word *symmetry*, the teacher introduced the "discrete symmetry machine." This sketch consisted of twelve coloured squares arranged symmetrically around a vertical line of symmetry. These squares move discretely on a square grid background. Dragging any square on one side of the line of symmetry will also move the corresponding square on the other side of the line of symmetry (see Figures 6.1a and 6.1b). The discrete motion, as well as the use of the grid, was intended to help the children attend to the distance between a square and the line of symmetry. The children were asked to observe and describe what they noticed about how the symmetry machine worked. They were invited to explain what will happen by coming up to the screen and showing their predictions. The teacher used the children's descriptions and predictions to emphasize the motion of the squares in relation to the line of symmetry (the square moves *along, away from* or *towards* the line).

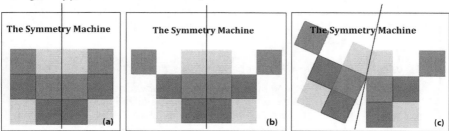

Figure 6.1: (a) The discrete symmetry machine; (b) After dragging one block away from the line; (c) After rotating the line of symmetry.

Using large cardboard diagrams of various symmetric and non-symmetric configurations (see Figures 6.2a–f), the teacher asked the children to re-create the designs or explain why the symmetry machine could not have created the design. With the latter designs (see Figures 6.2c and 6.2d), the children were encouraged to talk about the relationships that should be present in order for a design to be symmetric. For instance, when asked by the teacher to explain why a given design is not symmetric, a child walked up to the sketch as he said, "Well, I just want to say something about the picture, because, this orange one is under the purple, then this one should be under the purple too." As he talked about each of the orange and purple squares on both sides of the symmetry, he used his right pinky finger to touch each of the squares on the sketch.

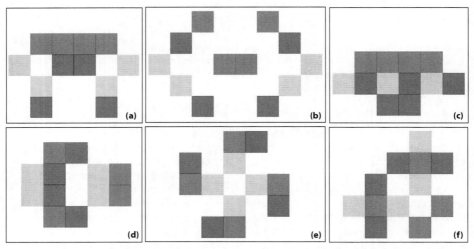

Figure 6.2: (a) Vertical symmetry; (b) Vertical symmetry; (c) Not symmetric;
(d) Not symmetric; (e) Horizontal symmetry; (f) Oblique symmetry.

The children were then asked to sit at one of the two large tables in the room and to make drawings of configurations that the symmetry machine could produce. While the children were working on their drawings, several children were asked to explain their drawings. The following excerpt shows one child's response.

Sarda's use of diagrams after the introduction of the symmetry machine

Sarda: If I move the orange upward [draws up arrows in Figure 6.3]; they both would. If I were to move the magenta down [draws left-hand down arrow], the other magenta will go down [draws right-hand down arrow]. If I move the blue sideways [draws left-pointing arrow], it will go opposite [draws right-pointing arrow].

Figure 6.3: Sarda's drawing during the first lesson.

The second lesson consisted of two parts. First, the children were invited to explore the discrete symmetry machine with a horizontal then oblique line of symmetry. The teacher encouraged the use of the same language developed in the first lesson so that the children describe how the squares move along, away or toward the line of symmetry. Then, a new sketch with only one side of the line of symmetry was shown; the children were told that the symmetry machine is broken and they must therefore predict what the other side of the line should look like. The sketch included different pages, each of which had a broken symmetry machine with a different type of line of symmetry (see Figure 6.4a–c).

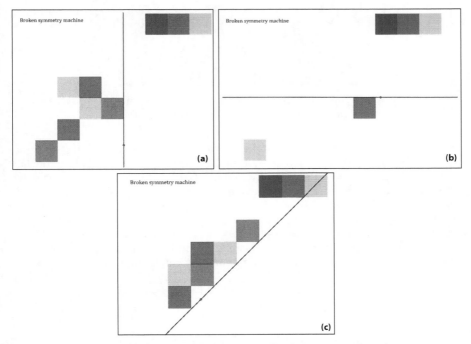

Figure 6.4: (a) Broken horizontal machine; (b) Broken vertical machine; (c) Broken oblique machine

On the second page, there was a large space between one of the squares and the line so that the students could not use the line as a visual reference. As can be seen in the transcript below, the students used their hands to mark out the "distance" from the square to the line:

Jess: [whispering] One, two [pointing at the empty space between the square and the horizontal line, see Figure 6.5a]. I will move the blue one, right over, up there.

Teacher: Up there, ok. I saw you were doing something like that [used her left index finger to point towards the empty spaces directly above the blue square]. What were you doing?

Jess: I was counting how many squares in the space.

Teacher: How many did you figure out?
Jess: Two.
Teacher: So where do you want me to put the other one?
Jess: Right on the top.
Teacher: Can you use the fact that this is two squares away to tell me where to put the other one?
Jess: It's two squares away from the line, so one, two, and three. [Figure 6.5b]

Figure 6.5: (a) "One, two." Jess counting the number of "squares in the space."
(b) "It's two squares away from the line, so one, two, and three."

Jess first whispered "one, two" to "count the number of squares in the space" between the line and the bottom blue square. Then, he explained that "it's two squares away from the line, so one, two, and three," while he points to the empty spaces three units above the line with his right index finger. His understanding of symmetry thus evolved from being able to discriminate between symmetric and asymmetric to a higher level of understanding that involved properties and reasoning about symmetry. This reasoning was evident in a subsequent paper task when the children were asked "Why is the square in the wrong place for a symmetry?" as seen in Sian's work:

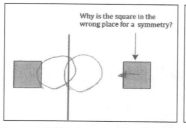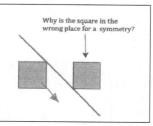

Figure 6.6. Sian's reasoning of "Why the square is in the wrong place?"

Figure 6.6 shows Sian's reasoning about symmetry using two different kinds of devices. First, the circles on the left figure are used to represent the spaces between the square on the left and the vertical line of symmetry. This helped Sian to reason that the squares are not equidistant from the line of symmetry. On the other hand, Sian's reasoning is characterized by the use of arrows. In both his drawings in Figure 6.6, Sian communicated the "wrong place" of the squares by indicating arrows showing his desired movement of the squares.

In the third lesson, the teacher introduced the "continuous symmetry machine" (Figure 6.7), and asked the children to observe and describe what happens when one point is dragged on the screen. Since there were only two dots on the screen, it was less obvious what the sketch was meant to do and whether or not vertical symmetry was involved. The teacher invited the children to draw particular symmetric shapes, which they had to identify before they began dragging. Every child in the class had the chance to come up and make a symmetric drawing. During this time, the children also noticed that at some locations on the sketch the two points would "get married," which initiated a class discussion about all the points where this might happen.

The students created a trace of the line along which the points converged and the teacher linked this line to the one on the discrete sketch.

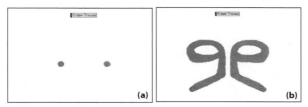

Figure 6.7: (a) The continuous symmetry machine;
(b) The trace left by dragging the point.

An interesting discussion unfolded when a child was asked to draw a house using the symmetry machine. He first drew the outline of a stereotypical house with a square base and a triangular roof. Then, he drew a tornado, a chimney and a door on one side of the house. This resulted in two tornadoes, two chimneys and two doors in the drawing (Figure 6.8b), which seemed unorthodox to his classmate. This prompted the teacher as to how he came to have two tornadoes.

14:50	Toni:	Because his markers, his pens, they are both going the same way. It's the same thing! One's on that side [raises his left hand]. One's on that side [raises his right hand; see Figure 6.8b]. They got to do the same thing!
15:13	Teacher:	Oh my goodness, a house with two chimneys!
	Students:	[laughter]
15:22	Teacher:	How come there are two chimneys?
	Students:	Two doors! It's opposite. It's opposite.

Figure 6.8: (a) Two tornadoes and a house with two chimneys and two doors;
(b) Toni raised both arms while explaining why there are two tornadoes.

The children in this episode are thinking dynamically about symmetry. Toni's explanation that the markers are "going the same way" accompanied by raising both arms up in the air shows his thinking that the tornadoes are drawn simultaneously and dynamically. The image of the house created by Toni opened up the possibilities for the teacher to question why there are pairs of tornadoes, chimneys and doors in the drawing. The children explained that the objects are "opposite," implying that they are two of each because they are on opposite sides of the line of symmetry.

Forms of spatial reasoning in the symmetry activities

In terms of the typology of spatial skills described in Chapter 2, symmetry at this age is often explored using visual identification that involves intrinsic-static skills. In these lessons, the children were also engaging in both intrinsic-dynamic and extrinsic-dynamic spatial reasoning and, more importantly, doing so within the same task. For the discrete symmetry machine, the children first made sense of how the symmetry machine was working through extrinsic-dynamic spatial reasoning and so noticed relationships between how one square would move in relation to another. By virtue of this work, they could then make predictions about which designs had been made by the symmetry machine, but now using intrinsic-static spatial reasoning as well. The importance of the extrinsic-dynamic spatial reasoning can be seen in the diagrams that the children drew, with the arrows indicating their coordination of the two different sides of the line of symmetry. This was eventually formalized in the broken symmetry tasks when the children began to attend to the equidistance relationship between two corresponding squares. During both the whole class interactions and the diagramming work, the children also engaged in intrinsic-dynamic spatial reasoning when they were introduced to the horizontal and, especially, the oblique line of symmetry. Several children turned their heads as they worked with the oblique line of symmetry so that they could interpret the motion of the squares in terms of vertical symmetry. Finally, with the continuous symmetry machine, we also see a back and forth between the extrinsic-dynamic and the intrinsic-static as the children must first decide how to make their target objects, which they have already identified as being symmetric (intrinsic-static), and to think about these objects in intrinsic-dynamic ways while they are dragging the point on the screen. But in the discussion of the chimney and tornado, there is a return to intrinsic-static.

Next steps

With the Grades 1–2 children, the idea of equidistance came out as an invariance in the dragging of the squares, and the children were able to use first visual cues, and then external objects (starting with their hands) to "measure" and compare distances to the line. However, it was more difficult for the children to articulate the perpendicular bisector relationship between the line of symmetry and two corresponding squares or points. This made it more challenging for them to work with oblique lines of symmetry. One way to mediate the development of this property of symmetry would be to introduce grid lines that would help the children see the way in which corresponding points line up. This could be a fruitful direction to take in a Grade 2 or 3 classroom. Once children have had experiences with symmetry, it may also be fruitful to investigate 2D shapes in terms of internal symmetry. Symmetry would provide an additional property that children could use to sort 2D shapes, and also to develop distinctions between, say, isosceles triangles and scalene ones.

Congruence and transformations: an exploration with pentominoes

The example we present in this section comes from work we (authors Bruce, Moss, Hawes, & Caswell) have been doing in the Math for Young Children (M4YC) research project; a professional development project for the enhancement of geometry and spatial reasoning in early years classrooms (Moss et al., in press). The research program has two related missions: first, to support early years educators and school administrators in gaining content knowledge of geometry and spatial reasoning for classroom use; and second, to design and field-test assessments, activities, and challenging but developmentally appropriate curricula for 4–8 year old students (Hawes et al., 2013; Hawes et al., in press). Amongst our outcomes is clear evidence that even the youngest students can engage with challenging geometry activities when provided with high quality opportunities to do so. Indeed, aligned with other scholars of early years mathematics, such as Clements, Ginsburg, van Oers, English, Mulligan, and Papic, our project reveals the kinds of sophisticated geometric reasoning young children can construct that typically would be attributed to older children (see for example, Bruce, Flynn, & Moss, 2013; Caswell et al., 2013; Moss et al., 2013).

In the following lesson, students were challenged to create and discover the twelve unique 2D shapes that can be composed with five unit squares connected so that at least one side is perfectly aligned, or flush with another – what we call the Pentominoes Challenge. In this task, the central focus was on congruence/non-congruence, rotations, and translations.

The lesson began with a narrative regarding helping the Princess Kate find twelve "magic keys" to unlock the prince from a tower. First, students were introduced to the idea of configurations of five same-colored squares with some restrictions such as "no gaps, no overlaps, no corner to corner, only flush sides" (Figure 6.9a).

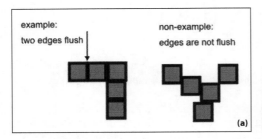
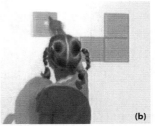

Figure 6.9: (a) The configuration on the left has flush edges with no gaps or overlaps; (b) Student dragged squares on the interactive whiteboard to construct one of the twelve pentomino figures while following the restrictions of the configurations. The teacher acknowledged that the student's shape was indeed a "magic key."

The teacher, Daniela, then went on to build two congruent pentomino figures on the interactive whiteboard, but in different orientations. She told the students that she was making two *different* magic keys.

Teacher: There, now I have a new one.
Students: [excitedly calling out] No it's the same! It's the same!

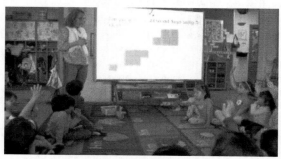

Figure 6.10: The teacher constructs two congruent figures in different orientations.

Teacher: [feigning confusion] What do you mean?
Sam: You can rotate it and it will be the same [moving his arms].
Teacher: [holding up a congruent plastic pentomino figure at yet another orientation] What about this one? Do you think this one is the same as well?
Students: [again with excitement] No it's the same too.
Students: [lots of voices] You just have to make it go the same way and you will see.

Pairs of students then worked together to find as many unique combinations of five squares as possible.

Figure 6.11: One student making different configurations of 5 squares.

Children were also encouraged to compare their shapes using five square tiles to pre-constructed pentomino figures, and find the "same" pentomino piece as they had created. Figure 6.12 shows a boy comparing one of his configurations with one of the pre-constructed pentominoes. Like many other children of his age, he was quick to recognize that the two pentominoes were congruent, but struggled to prove it through overlaying the solid pre-constructed pentomino onto his own. After some productive struggle, the boy realized the importance of "flipping" the shapes to prove congruence. Having being provided sufficient time to build at least three of their own configurations, children worked in small groups to share their configurations. Each small group then worked together to identify and prove all unique configurations from amongst the groups' collection. More time was then provided for the groups to work toward finding any of the remaining pentominoes – better known to the children as the "magic keys." As part of this activity, most of the children named their shapes as a letter of the alphabet or a common object.

Student: [holding up the shape of a cross] It's a plus.
Teacher: Is there a different way you might describe that, Cecil and Lana?
Cecil: It's like an x for a pirate treasure, or, it could like – turn into a box!
Teacher: How could you turn it into a box?
Sam: But it's a box without a top. You lift these sides up, but you need one more … one more on top to get a cube.

As an extension that drew on student curiosity, the collaborating teachers asked students to look for the seven figures in the pentomino set that could be folded to make "boxes without tops." Students were engaged for long periods, and were fascinated by this investigation.

Figure 6.12: Comparing square configurations to the pre-constructed pentomino shapes for congruence.

Figures 6.13: Visualizing and paper folding to determine which pentomino configurations make an open box.

Although the mathematics of pentominoes (and other polyominoes) is usually reserved for much older students, each time we introduced pentominoes to kindergarten students in our research classrooms, we found the same level of sophisticated reasoning and deep engagement (Naqvi et al., 2013). We also noted the kinds of geometric reasoning that the students engaged in as they participated. The students naturally used visualization to assess whether shapes were congruent and were able to both mentally and physically rotate and reflect the shapes as well as recognize and represent shapes from different perspectives. This kind of reasoning is not typically expected of such young children. Surprisingly, these 5-year-old students were not only able to assemble and distinguish each of the twelve pentomino shapes; they also initiated, with the help of one of their classmates, the dynamic connection between 2D and 3D figures through first mentally and then physically folding the nets to make "boxes without tops."

Engaging the mathematical imagination through mental rotation activities

The imagination plays a central role in mathematics, providing the necessary mental space to playfully create, manipulate, and carry out mathematical ideas, problems, and proofs. For over 40 years, psychologists have used mental rotation tests (see Chapter 2) as an attempt to access the "mind's eye" and gain insights into the human ability to imagine the movement and transformations of 2D and 3D objects. Since the seminal work of Shepard and Metzler (1971), mental rotation has emerged as *the* prototypical measure of spatial reasoning. Individual differences in the skill have been linked to performance across a variety of activities, including success in science and, perhaps most notably, mathematics. Research shows that people who perform well on measures of mental rotation also tend to perform well across a range of mathematical tasks, including arithmetic, algebra, calculus, geometry, and word problems.

What explains the relationship between mental rotation skills and mathematics performance? What are the pedagogical implications of this relationship? We believe that the answer to both of these questions is in part linked to the role of the imagination. In performing mathematics and mental rotations, we must use our imagination to mentally manipulate information. While mathematics commonly involves the mental manipulation of symbols, mental rotation involves the mental manipulation of 2D and 3D objects. It is thus reasonable to consider how mental rotation skills and mathematics are connected through their shared involvement of "mental manipulations" – a key feature of the imagination.

In terms of pedagogical implications, we believe that increased efforts are needed to explicitly engage the "mathematical imaginations" of young children. We see mental rotation as assuming an important place in early years mathematics curricula. Not only are young children capable of considering complex and abstract mathematical ideas (e.g., infinity and zero) but are also at an age where they can begin to imagine the movement and transformations of 2D and 3D objects. By designing curricula that engages children in developing their mental rotation skills, we are moving towards a more dynamic and imaginative approach to early geometry instruction and learning.

As part of the collaborative work of educators and researchers in the Math for Young Children research program, we have created a series of lessons and activities that aim to help young children develop their mental rotation skills. In the following section, we describe a Grade 1 lesson that involves visualization and mental rotation. We then briefly describe two other activities that target mental rotation and have been field-tested with children in Grades K–3. We conclude this section by discussing the role that mental rotation played in the described lessons, but also the role of mental rotation in early years mathematics more generally.

Upside Down World: a Grade 1 lesson on visualization and mental rotations

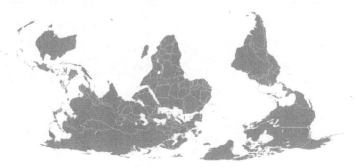

Figure 6.14: "Upside down world map."[1]

Imagine the world flipped upside down. Flipping a world map "upside down" from its typical orientation offers a new spatial perspective. It forces us to reconsider spatial relationships both within and between geographical objects. Australia is no longer tucked away but assumes a commanding position. Russia no longer perches on top of Eurasia but provides the continent's backbone. Transforming visual-spatial information not only helps gain new insights or "ways of seeing" but also figures prominently in STEM thinking, including architecture. As shown in Figure 6.15 (next page), the architect Gaudí constructed upside down models of buildings, using gravity as a means to build catenary arches. Later, the models were flipped "upright" and transformed into freestanding blueprints.

In the following Grade 1 lesson,[2] we describe one Lesson Study team's attempts to foster this type of "upside down" thinking. With the specific goal of targeting visualization and mental rotation skills, the team of teachers and researchers designed a lesson that moved children through a series of challenges where they used multilink cubes to reconstruct a fictional "upside-down world."

Lesson description

As a warm-up to the lesson, children were invited to engage in a short activity designed and named "Building with the Mind's Eye." Farah, the first grade teacher, asked that her children join her in a circle. She reminded her students that during the past weeks they had been doing much building with cubes. *"We're now going to try something different,"* she told her students. *"We're going to try building in our minds, with no cubes at all."* Intrigued by the challenge,

Figure 6.15: The famous Spanish architect Antoni Gaudí (1852–1926) used chains, weights, and gravity to build catenary arches (see image on the left). By suspending weighted chains and attaching them from a ceiling board at various distances from one another, Gaudí used gravity in his favor to create parabola-like arches. Flipping the structure upside down provided the inspirational blueprints for structures such as the Church of Colònia Güell featured on the right.[3]

her students enthusiastically responded to the request to close their eyes and listen carefully. In a gentle tone and at slow and deliberate pace, Farah delivered the following instructions:

> Imagine that you have three blue cubes and one green cube. Now, take the three blue cubes and snap them together to form a tower. Lay your blue tower on its side so that it's now lying on the ground. Take your one green cube and attach it on top of the middle blue cube.

> Think really carefully about what your figure looks like because we're now going to do something special to it. Here's the trick! Flip your figure upside down! Now, what does the picture in your mind look like?

Students were then presented with a picture of three adjacent cube figures composed of blue and green cubes (see Figure 6.16). One by one, Farah pointed to each figure and asked her students to applaud their approval of whether or not the figure matched her description. When asked how they were able to identify the correct match, students demonstrated deductive reasoning: "*It can't be that one* (pointing to the referent), *because it has three green cubes and you said three blue cubes … not three green cubes.*" One student referred directly to the rotational aspect of the task, stating that she had imagined an upside down "T" being transformed into its upright position. Having captured her students' interests, Farah transitioned seamlessly into the main lesson involving the following fictional narrative around 3D mental rotation.

"Imagine a world made entirely of cubes," Farah began. A brief description of the "cube world" and students' interests were piqued. "Even the people?" asked one boy. "Yes, even the people," responded Farah. At this point, a series of figures constructed with interlocking cubes were revealed

Figure 6.16: Farah asking her students to identify the structure
they had "pictured" in their imagination.

and placed in the middle of the carpet area and were said to represent
buildings. As the narrative continued, students were introduced to the
protagonist, George, a curious city worker faced with a major temptation:
a big red button located in the city center. Every day he passed the button
on the way to work and wondered what would happen if he were to press
the forbidden button. Unable to resist the temptation any longer, George
presses the button and with that, the world gets flipped upside down. Farah
mimicked the event by flipping each of the buildings placed in the middle
of the circle. Students were asked to imagine various scenarios implicating
visual-spatial reasoning, for example, "What would the shape of the CN
Tower now look like?" Students were quick to join in with their own probing
questions, "How would you get into buildings? The door would be way up
in the air." "What would happen to all of the oceans ... and what about the
sky ... would they be switched?" After this brief and imaginative discussion,
Farah returned to the narrative and explained that George needed their help.
He needed help re-constructing his city, a feat that would require reverse
engineering and strong visual-spatial problem-solving skills – not unlike the
creative thinking of Gaudí!

To begin, Farah asked her students to look carefully at the upside down
structures she placed in the center of the circle (see Figure 6.17, next page).
Students were then given individual bags of interlocking cubes and invited
to work in pairs, one student as the role of builder and one student as the role
of instructor. In turn, students described to their partner how to re-build the
structures from the base up. Those listening to the instructions built accordingly.
This exercise provided students with the opportunity to exercise both their
expressive and receptive spatial language skills. Upon the completion of each
building, students compared theirs to the upside down buildings in the middle
of the circle. After re-building the complete set of structures, Farah revisited
the narrative and in a dramatic fashion flipped the structures right side up
again, providing students with an opportunity to compare their structures
with those in the middle. Students were congratulated for helping George
re-build his city.

Throughout this lesson, students were engaged in a number of dynamic
spatial and geometric skills, including visualization and mental rotations

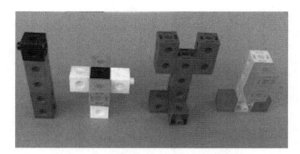

Figure 6.17: Examples of the structures students were asked to
reverse engineer and re-build "right side up" again.

through the composition and decomposition of 3D shapes. During the warm-up activity, children were faced with the difficult task of building the cube-structure mentally and then rotating the figure by 180°. When presented with a picture of three figures and asked to indicate which figure they had built and mentally rotated, the majority of children correctly identified the described figure. This is an impressive feat, and a testament to young children's capacity to engage in complex spatial reasoning tasks. The task of re-building the "upside down world" challenged students to imagine what the buildings would look like if flipped upside down (i.e., mentally rotating the buildings). Being able to see and describe the top of the building as its base proved difficult for most students, but was not out of reach. Instead, all children appeared engaged with the task, and worked collaboratively to complete the task. When the instructions did not achieve the intended result, students made it known and helped indicate where and why the building instructions broke down.

Other activities developed to increase children's mental rotation skills

Cube challenge: building unique 3D structures

In this activity, students were challenged to build unique 3D structures using sets of 3, 4, and 5 multilink cubes. To begin with, students built using only sets of 3 cubes. Although only two unique configurations are possible, children often grappled with the question of invariance. The equivalent configurations among their constructions inspired students to ask themselves whether the two structures are the "same" or "different." In our observations, most young children typically viewed two equivalent structures as different so long as they were presented in different orientations. It was only after a class-wide discussion or debate amongst peers that children came to recognize equivalence in 3D shapes of different orientations: "If you just turn this shape over, it's the same." After building with 3 cubes – and it was clear that students understood the principle of invariance – children were asked to build unique shapes using sets of 4 cubes and eventually 5 cubes (see Figure 6.18 for all possible unique combinations using 4 cubes).

In introducing students to the 4- and 5-cube challenge, students were invariably confronted with mirror images. For example, the two far right configurations in Figure 6.18 are mirror images of one another, and for this reason are unique in structure. In the words of one Grade 2 student, "*No matter what you do to this one* (physically rotating one of the last two figures), *you can not make it look the same as this one. They're different."* According to a Grade 1 student, 3D mirror shapes are different because *"the only way to make them the same is by moving one of their parts."* Being able to recognize and understand mirror images as unique structures is challenging (even for adults) and constitutes the central difficulty of this activity.

Figure 6.18: Unique configurations that can be built using sets of 4 cubes.

Throughout this activity, there are primarily two methods to determine whether two or more shapes are unique or equivalent. The first method is through direct comparisons made possible by physically rotating and aligning the structures being compared. The second method involves visual comparisons, including the act of mental rotation. For example, to determine that the 3D shapes in Figure 6.19 are equivalent, one strategy is to mentally rotate one of the figures 180° around the vertical axis. This mental rotation strategy can be encouraged and is a necessary form of proof when students are not able to directly compare shapes (as might be the case when students are comparing their shapes to those of their peers). To encourage mental rotation strategies, the teacher can ask students to sit in a circle and place some of their configurations in the middle for everyone to view. The teacher can then hold up one of the structures and have students compare this shape to the others to determine 3D equivalence or uniqueness.

Figure 6.19: Example of equivalent 3D configurations.

Shape transformer

In this activity, students were introduced to a "machine" called the "Shape Transformer." As shown in Figure 6.20, the machine had both an input and output slot. The teacher began by explaining that the machine has special powers and that anything that goes into the machine is transformed and comes out as something different. To demonstrate, the teacher presented students with an input card and then inserted the card into the appropriate "input" slot. Another teacher (or student) played the role of the machine and slid the corresponding "output" card through the output slot, making appropriate machine-like noises. The teacher then placed the input and output cards next to one another and asked students to carefully observe the relationship between the cards. After several more examples, students were encouraged to try to guess the machine's rule. Using mini whiteboards, students were presented with an input card and then asked to draw what they predicted the output card would look like. After students had successfully solved and described (or drawn) the machine's rule, the teacher reconfigured the machine and began a new series of input and output challenges according to a different rule. Eventually, students created their own series of input and output cards and presented their challenges to classmates in a "guess my rule" game.

This activity offered a playful opportunity to engage young children in a precise and focused spatial reasoning challenge. This activity proved to be an especially effective tool for introducing children to problems that involve geometric transformations, including those that involve mental rotation. Children were motivated to discover the machine's rule and in doing so, learned to pay careful attention to the spatial relations between the input and output shapes. In explaining the machine's rule, children were provided with opportunities to use spatial language to describe the geometrical transformations.

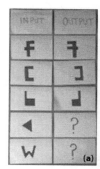 

Figure 6.20: (a) Example of input and output cards: the rule for this example is that of a reflection or a 180° rotation about the vertical axis; (b) A team of teachers discuss (and gesture) the solution to a "shape transformer" challenge that involved a rotation.

Summary

In the lesson and activities described above we gained insights into how children can develop their spatial reasoning through classroom-based activities that explicitly target mental rotations and that elicit high levels of engagement. These activities depart from the typical approach to early years geometry instruction, in which children deal with primarily static features of geometry including the labeling and sorting of shapes, and rather, invite children to imagine the movement of shapes and objects. When children engage in mental rotation activities they develop what we call "mathematical imagination"; that is, the ability to playfully generate, retain, and manipulate objects and symbols.

The role of touchscreens in fostering spatial reasoning

Each of the vignettes presented in this chapter illustrates how multitouch technologies and other tools supported dynamic learning environments for young children, where the children visualized movement in space, discussed the effects of movement on objects in space, and enacted these movements. One structure some teachers have adopted for these types of spatial reasoning is called the "Three V's": Visualize, Verbalize, and Verify. The students are encouraged first to visualize what is happening (given the task at hand), then to discuss it with others in the class, and finally to verify what they think is occurring by taking action (enacting the movement or change on an object and comparing this to its original iteration or other iterations). As part of the classroom-based interventions described in the vignettes, touchscreens often acted as a mediator for the development of spatial reasoning. As a result, these interventions have engendered a relatively novel area of research regarding the impact of touch technologies on children's spatial reasoning. One key aspect to these new research explorations involves investigating how to capitalize on eye–hand relationships through touch gestures (Sinclair & de Freitas, in press) for meaning making.

The use of gesture for meaning making

Alibali and Nathan (2007) define *representational gestures* as those where "the hand shape or motion trajectory of the hand or arm represent[s] some object, action, concept or relation" (in Nathan, Eilam, & Kim, 2007, p. 536). These gestures have been strongly linked to spatial skills such as mental rotations (Frick et al., 2009; Chu & Kita, 2011). Cook, Mitchell, and Goldin-Meadow (2008) found lasting effects (after four weeks) were greater for students who were encouraged to gesture (85% retention rate) during equation balancing than for those who did not gesture (33% retention rate). In a similar study with adults, Chu and Kita (2011) found that gesturing students were able to perform more complex mental rotations. These studies consider the use of gesture as hand/body movements "in the air" that represent or communicate thinking.

Multitouch screens on the other hand are tactile or haptic in nature. That is, they cause material effects (Roth, 2001; Radford, 2002; Nemirovsky, Kelton, & Rhodehamel, 2013), and this phenomenon has caused a substantial reconsideration of the relationship between the encounter (touching the screen) and the representation (what appears from the touch gesture) (Sinclair & de Freitas, in press), and further – a significant reconceptualizing of the range of gesture types, their nature and their impact.

> ... the hand actually operates very close to the surface of a screen: pointing to objects on the screen by tapping them; sliding objects along on the screen so as to leave visual and aural traces of the finger's path; pinching objects together in order to make new ones. These gestures of pinching and flicking and pointing both communicate meaning and inscribe marks. (Sinclair & de Freitas, in press)

Some current spatial reasoning research in mathematics education is exploring touch gesture as a particularly powerful affordance that offers children (and other users) unique ways to engage in spatial reasoning. As an example of extending spatial reasoning activities further through touchscreens, author Bruce generated an iBook called the "Four-Cube Challenge" (available for download at www.tmerc.ca/m4yc). The iBook invites children to use real interlocking cubes to make all possible combinations of four joined cubes (see vignette above for more information about this task). Then, when verifying (to see if they have found the comprehensive set of configurations), children are encouraged to compare their figures to the pre-generated configurations in the storybook on the iPad screen. Of particular importance here is that the 3D shapes in the iBook can be rotated using a directional tracing touch gesture of the finger in the circular direction in which the user wants to spin the shape. The child can also use two fingers to "pick up" the digital shape and expand its size (proportionately) and its orientation. Both of these touch gestures cause the shapes to rotate on the screen, in response. The shapes rotate in every direction fluidly about a fixed internal central point.

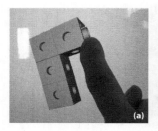 

Figure 6.21: (a) Pointing-tracing and (b) iconic-whole object touch gestures used to rotate the virtual 4-cube configurations.

These rotating touch gestures enable visual comparing of concrete interlocking cube figures generated by the students, to screen figures by aligning the orientation of real and pictorial images, but they also enable the comparison of multiple figures on the iPad for equivalence. The power of the visual figures, combined with their touchscreen motion properties pushes children to consider the similarities and differences of the two mirror configurations in the set of solutions, in particular (Figure 6.22).

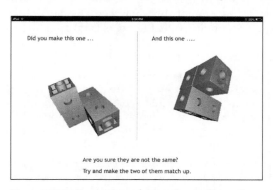

Figure 6.22: Mirror image four-cube configurations.

Author Sinclair's TouchCounts™ iPad application[4] also supports meaning making that is mediated through touch gesture, in this case bringing spatial reasoning, digit kinesthetics and quantity together for users. In this application, the child performs actions with her/his fingers that embody number concepts. For example a tap with four fingers results in four circles being generated on the screen, encompassed in a larger circle with the symbol "4" at the center. A two finger pinching gesture brings circles of quantities together in order to add them – if the child has three dots in one circle and four dots in a second circle, and then pinches these together (see Figure 6.23c), the sets are joined to make one circle of seven dots.

Figure 6.23: TouchCounts gestures of fingers tapping for quantity and pinching for addition.

The pinching touch gesture offers a way of expressing the concept of adding, calling on metaphors from the spatial domain involving the gathering of things together. This TouchCounts example provokes us to think more deeply about how facets of mathematics reasoning cluster into a repertoire of gestural kinesthetic spatial visual and conceptual representation combinations through the use of the App. Compared to handling physical tools, the use of virtual tools on touchscreens provides students with *different* kinesthetic experiences (Ehrlich et al., 2006). Thus, given that input on an iPad directly employs one's fingers, the potentially unique role of touch gestures within the interactive environment of tablet technologies is an area of research worthy of greater attention.

Linking ideas

The examples presented in this chapter feature specific tasks, carefully selected materials, and an enhanced mathematics curriculum that capitalizes on, and cultivates, spatial reasoning in playful, engaging and mathematically rigorous ways. The tasks move well beyond typical classroom mathematics activities and lessons in early years classrooms, and in particular encourage dynamic types of spatial reasoning. Results of our intervention studies (through which the above activities and many others were generated) point to the power of spatializing the curriculum. For example, assessments conducted before and after introducing children to intervention activities including the ones above, have revealed significant growth in students' spatial language, geometrical reasoning, mental rotation and numerical comparisons (Bruce & Hawes, in press; Hawes et al., in preparation). Moreover, and importantly, children have demonstrated high levels of engagement and persistence with these activities.

In our research with young children and early years teachers, the very act of learning about spatial reasoning and coming to understand its importance, and collaboratively designing tasks has had positive outcomes for students, teachers and researchers. The broader implication here is that curriculum reform in geometry and related classroom learning opportunities are not only possible, but necessary. Part of this reform movement involves educators and researchers working together to generate more high quality examples of how to engage children in dynamic spatial reasoning activities in classroom contexts.

Notes

1   Source: http://commons.wikimedia.org/wiki/File:Blank-map-world-south-up.png

2   Source of catenary image: http://en.wikipedia.org/wiki/File:Maqueta_funicular.jpg; source of Church of Colònia Güell image: http://en.wikipedia.org/wiki/File:Cripta_de_la_Colònia_Güell_1.jpg

3   To view a brief video of the lesson, see: http://www.oise.utoronto.ca/robertson/

4   Available at the App Store.

7

Motion and markings

LYNN McGARVEY, NATHALIE SINCLAIR, JENNIFER S. THOM, DONNA KOTSOPOULOS

In brief …

This chapter focuses on the dynamic dimension of spatial reasoning, particularly in terms of how drawing is used to develop and communicate mathematical ideas. Drawing is a way in which children explore and become aware of spatial concepts and relations in their dynamic, embodied and 3D world, and also a way for them to record and reflect on that motion.

Motion in the mathematics curriculum

Recognizing the fundamentally kinesthetic experiences of humans-in-the-world, scholars have suggested replacing the old Cartesian adage of "I think, therefore I am" with the more embodied one of "I move, therefore I am" (see Chapter 5). Indeed, in our temporal world, change is constant, albeit often invisible. Even looking at an image, which can be thought of as static, involves change as our eyes scan its contours and perceive various hues and shades – we ourselves change as we gaze.

However, motion has always been a very troubling phenomenon in mathematics. For example, the ancient Greek philosopher Zeno pointed to the paradox involved in moving from one position to another by repeatedly going halfway – and never arriving at our destination. Plato derided ancient geometers who talked of points moving and triangles being superimposed because, for him, mathematical objects were immanent, safe from the vagaries of the temporal and mobile nature of human existence. Even two millennia later, we find the mathematician Bertrand Russell (1903/2010) expressing suspicion: "To speak of motion implies that our triangles are not spatial, but material. For a point of space is a position, and can no more change its position than a leopard can change its spots" (p. 405). Indeed, as Balacheff (1988) has observed, formal mathematics is characterized by "three d's": depersonalization, decontextualization and detemporalization. In this formal mathematical discourse, mathematical objects have no histories, no movement, and they exist forever.

Recently though, cognitive scientists such as Lakoff and Núñez (2000) have identified significant sensorimotor experiences that shape our mathematical understanding, such as "motion along a path," which acts as a metaphor that enables us to make sense of addition, for example, as the end point of path lengths. Experiences like these are fundamentally mobile and provide important bases on which we can make sense of mathematical ideas. Further, from his study of mathematicians' lectures (Núñez, 2003) and from their study of mathematicians' descriptions of mathematical concepts (Sinclair & Gol Tabaghi, 2010), it is evident that mathematicians think of mathematical objects such as limits and eigenvectors in highly dynamic ways. Indeed, the mathematician William Thurston (1995) includes "process and time" and "vision, spatial sense, kinaesthetic (motion) sense" as two of the six major facilities that are important for mathematical thinking (pp. 4–5). In other words, while formal mathematics may well yearn to be atemporal, it seems that learning and doing mathematics is not.

In this chapter we focus on the aspect of spatial reasoning that involves motion. We are interested in the ways that young children might use motion to help them do mathematics across the curriculum, whether it involves moving in space to orient themselves, moving triangles to explore their various possible shapes or moving numbers to explore operations such as addition and subtraction. The recent developments in embodied cognition that were described in Chapter 5 point to ways in which a moving body – a gesture in the air or on the screen, a turning of the head to identify a symmetry – may come to know mathematics. At the same time, recent developments in digital technologies are providing access to motion that enable students and teacher to reason in ways that previous pencil-and-paper technologies have not.

While educators are increasingly recognizing the importance of tactile experiences, kinesthetic bodily movement, and dynamic shape transformation to support the learning of geometric concepts, the very nature of these time-sequenced activities make them fleeting. They leave only a bodily trace with no record for later reflection or discussion. In recent years, the development of digital tools makes such traces possible. For example, a video or photo-series can capture the bodily or manipulative movement of children and computers can also record and replay sequences of actions of dynamic objects. Yet, at present, these traces are not common or frequent in the day-to-day activities of children in classrooms. Indeed, the most accessible way for young children to record their ideas and actions is through their own mark-making or drawings. This is also true of mathematicians and, as the philosopher and historian of mathematics Gilles Châtelet (2000) has argued, drawings provide an essential link between the moving, gesturing body and the formal mathematics. Such drawings are not just representations of already-formed thoughts; instead, they allow for the creation of new ideas – they are the thinking. Using Châtelet's ideas, de Freitas & Sinclair (2012) have shown how students' diagramming of an animated film about circles enabled them to develop new ideas about how circles grow and eventually become lines.

In this chapter, we offer three vignettes in which children experience and explore motion in dynamic geometric contexts. Then, through their drawings we see further exploration and ways in which children begin to pictorially and symbolically make records of motion they have seen and imagined, and come to new understandings through their drawings. Prior to sharing the vignettes, an introduction to the role of drawing in spatial reasoning is necessary.

The role of drawing in spatial reasoning

Children's drawings have served several different purposes in research related to the teaching and learning of spatial-geometric concepts. Most prominent, drawing tasks have frequently been used in developmental psychology as a means to access and analyze children's cognitive development of spatial structures (e.g., Piaget & Inhelder, 1967). From this theoretical orientation, children's drawings have been used to assess their knowledge and awareness of spatial relations in and between shapes and objects in embedded figures (Kirkwood et al., 2001), spatial structures (Mulligan & Mitchelmore, 2009), and 3D scenes (Case et al., 1996).

Parallel research from a sociocultural perspective suggests that differences in children's drawings of shapes, 3D images, and problem-solving tasks are not necessarily attributable to normative cognitive development, but to cultural conventions and personal experiences (Cohn, 2014). What a child draws and how he or she draws it is often based on the cultural models and conventions available rather than internal schemas. Along these lines, studies have demonstrated that purposeful attention to the development of a visual vocabulary of lines, patterns, and 2D and 3D shapes through multiple experiences, including drawing, supports children's awareness of spatial images along with tools to express, describe, and produce them (e.g., Hershkowitz, Parzysz, & van Dormolen, 1996).

The ways in which children's drawings have been used in previous research have been to assess cognitive development and support learning; however, we have noticed a dearth in the literature whereby children resolve ways in which to enact, describe, and symbolize motion through drawings. We recognize that one reason for this neglect is due to the nature of drawing itself. Once an image has been drawn it becomes a static artifact that no longer retains the mobility that went into its making (the moving hand) and often objectifies the phenomenon to which it refers (Thom & McGarvey, in press). This is evident in graphs of the Cartesian coordinate system, which, despite the fact that they are meant to describe movement (speed and acceleration) are often interpreted by teachers and students as static objects.

In the three vignettes below, drawing is considered both a way in which children explore and become aware of spatial concepts and relations in their dynamic, embodied, and 3D world, and also a way to record and reflect on that motion. While mathematicians have devised highly sophisticated ways of doing this (like the Cartesian coordinate system), there are a variety of other techniques that are used in art and science (Cutting, 2002). For example, in a middle school setting, Sinclair and Armstrong (2011) show how students used a variety of cartooning techniques such as frames and action lines in their drawings of a story involving a character walking, stopping and running from a starting point to a destination. With the help of the teacher, these techniques were then associated with conventions of the Cartesian coordinate system in order to help the students better understand these conventions and connect them back to the motion in the story.

What follows are three vignettes on children's exploration and expressions of motion in geometric curriculum contexts through actions, drawings, verbalizations and gestures. In the first, children enact the motion of their physical bodies in a relay race through drawing and description. In the second, we share how children experience and produce dynamic motion of triangles first through dynamic geometric software and then through drawing. The final vignette involves the children's imagined navigation through a large scale-mapping project. In each case, we see how children attend to and reason through expressions of bodily motion occurring through time and through space. Their actions, the act of drawing, the products of their drawings, and the descriptions of drawings offer multiple ways with and in which children explore motion.

Three vignettes exploring motion in mathematics

Exploring and expressing motion: the relay race in junior kindergarten

In a morning and afternoon junior kindergarten class with a total of 31 3- to 5-year-olds, the teacher explored spatial relationships with children through children's literature, music, and photographs. Of particular interest were the ways in which the children explored motion through their actions, drawings, words, and gestures. The whole class activity was a regularly scheduled time where the teacher planned for movement through action songs or games such as follow-the-leader and duck-duck-goose. On this occasion, the teacher sat with the children on the carpet to prepare for a relay race. In addition to a discussion about children's experiences with relay races, the teacher and the children turned their classroom into a "racetrack." The start line for each team was placed at one end of the classroom. At the start were two baskets, each filled with several beanbags. At the other end of the room were two hula hoops – one for each team. The children were grouped into two teams and each sat on a cushion along either side of the racetrack. The race involved the children taking a beanbag from the basket, running to the end of the room, placing their beanbag in the hoop, and running back to their cushion. Once sitting, the next person on their team could go – repeating the same loop.

After several children demonstrated the path they would take, the children excitedly took their positions and the race began. The children cheered for their team, squirmed on their cushions, and took turns running the racetrack. They played two rounds of the relay race and then it was time to stop. The children were still excited and wanted to play again. Instead, the teacher promised to do the relay race another time and encouraged the children to draw themselves running the relay race. In the morning and afternoon classes, 23 children drew 34 pictures. We describe Caley's and Avery's drawings below followed by a discussion of Peter's and Surrey's pictures.

Caley's and Avery's relay race drawings

Caley and Avery are two of the 23 children in the morning and afternoon classes who drew their experiences in the relay race (see Figure 7.1). Both Caley's and Avery's drawings reveal some of the common conventions used by preschoolers. They used a multi-point perspective such as the cushions drawn from above, and human figures from the side; human figures are drawn as "tadpoles" where no distinction is made between the head and the body; and circles are drawn as closed figures, but not necessarily circular figures. In this discussion, however, we are interested in how the children use drawings to explore and describe the motion they experienced in the relay races.

Figure 7.1: (a) Caley's side-view drawing;
(b) Avery's head-on drawing.

Caley starts by drawing the running path as a horizontal line through the middle of the page from left to right. She makes another horizontal line above, this time from right to left showing the path back to the cushions. At the right end of the lower path she makes a small mark for the hula hoop. From here, Caley draws several images of children running along the path. The final part of her drawing actions includes adding cushions to the top and then the bottom edges of the image.

When asked to describe her drawing, Caley's gestures and verbal description involve a re-enactment of the relay race with her fingers as she sees the relay race taking place on the page. She says, "everyone is running" and gestures with her finger down the path from left to right. "You put a beanbag *in* the hula hoop and go back," tracing her finger horizontally across the top line from right to left. "Then you jump *on* the cushion." When asked if the children are all running at the same time, she uses her finger to gesture multiple loops around the path suggesting that the drawing is of several children all on the same team running one at a time.

Caley is the only child who drew the relay race from a side view where the movement is side to side on the page. The spatial words that she uses, including "in" and "on" are often known as static relations. That is, a beanbag can be *in* a hoop, and a child can sit *on* a cushion; yet, when Caley tells this story, both in and on are bodily action words. Another point of interest in her drawing are the multiple images of children simultaneously on the path, and yet her gestures indicate that she is using a series of discrete static images to signal motion of multiple moving objects. Her drawing is not static, but a storyboard that comes to life through gesture and story in her retelling of the event. While looking at the drawn image, it is important not to forget the temporal nature of the act of drawing; that is, the movements of Caley's hand as she made the marks – the track literally running its course as she draws the lines.

Avery, as did many other children, draws the running path starting at the bottom of the page and going to the top, and uses the convention of an arrow to indicate the direction. On either side of the line at the top of the page, he draws a hula hoop for each team with several beanbags inside each. He draws the basket of beanbags at the start line. On the right he draws a number of cushions as closed figures. Afterwards he pauses to count them by touching inside each circle leaving a dot. He shifts to the left side and draws five more closed figures counting quietly as he draws, "one, two, three, four," then checks again, "one, two, three, four" followed by one more closed figure, "five." He draws one human figure on either side of the vertical line half way up the page. He then adds two more human figures to each side of the line. He says he is "done" and the teacher sits down beside him to ask about his drawing. Before he describes his drawing he adds two curved lines with directional arrows to the left and right of the vertical line. He says, "This shows where to go in the relay." When asked about the human figures, Avery points to the figures on the left and says, "This is me." On the right he gestures, "This is Kieran. He is running." He points to the figure at the top left and says, "I'm going faster than Kieran." When asked about other features on his drawing he points out the "beanbags at the start," the "hoops with beanbags," and "these are the cushions we sit on to wait."

Similar to Caley, Avery draws multiple images to symbolize movement over time. Of particular interest is that his drawings indicate both direction through the use of arrows and speed of motion. Both the images of himself and Kieran midway up the page are relatively even, but his drawings of himself on the left are further down the track compared to his images of Kieran on the right.

Children's drawings of motion

There were 14 children who attended the morning class and 17 attended in the afternoon. Not every child participated in the drawings, but the 23 children who did produced 34 drawings. Video of the children's drawing activities captured only a few of the drawing actions and utterances of the children; however, each child had an opportunity to share his or her drawings with an adult. Of the 34 drawings made, 19 drawings, including Caley's and Avery's above, depicted a full-scale view of the relay race, revealing an awareness of the spatial relations of the mapping of the relay race elements. These included the starting line, the hula hoop, and many of them included the cushions along the sides of the track.

The children used multiple means by which to explore, symbolize, and describe the layout and movement they experienced through their drawings, descriptions, and gestures. A common means that the children used to evoke motion were similar to methods used in art, mathematics, and science including stroboscopic sequences of multiple stationary images along a path, and vectors in the form of arrows (Cutting, 2002). In other drawings, such as Peter's drawing below (Figure 7.2a), the movement is only apparent in the telling of the relay race. He says, "That's the straight line to run. We could draw lots of people there." In the middle, he has marked an X and says, "This is the guy in the relay race." Surrey's drawing does not include a mapping of the entire race. Instead, he describes a key action: "This is me dropping the beanbag in the hoop."

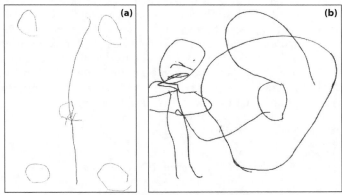

Figure 7.2: (a) Peter's drawing of the race;
(b) Surrey's drawing putting a beanbag in a hoop.

The action of drawing and the drawings themselves contribute to the experiences, conventions, and symbols to describe motion. The pictorial elements in the drawings are modified by the children using symbolic elements to describe direction, distance, and speed (Sherin, 2001).

Digital dynamic diagrams

In this next vignette, we see Grade 2 students engaged in depicting the temporal and dynamic motion of a shape in a computer environment through the static constraints of paper and pencil. In an hour-long session, students were introduced to the concept of a triangle along with special kinds of triangles including equilateral and isosceles. With the children sitting on the carpet, gathered around a screen on which a dynamic triangle was projected, the teacher led a conversation about the triangles that *Sketchpad*, a mathematics visualization software, could make using the dragging feature.

The discussion began, as expected, with most children identifying a shape as a triangle if it looked like the prototypical image of an upright equilateral triangle. Eventually, the children began to shift their discourse on triangles and used the word "triangle" as a family name to describe, not only the prototypical shape, but also upside down and skinny triangles. Further, they began to talk about the invariances of a triangle, noting that no matter how the *Sketchpad* triangle was dragged, it always had three sides and three vertices. This shift did not occur immediately and many children held fast to prototypical notions that triangles had to be, for example, "right side up." Following this whole class conversation, the children were asked to work in pairs with iPads running *Sketchpad Explorer*. They were given the task of using two or more triangles (already present in the sketch) to form a square. The goal was to allow the students to experience the dragging of the triangles on their own, and to encourage them to produce triangles of different shapes and sizes.

In a whole classroom setting afterwards, the teacher invited the children to show the different strategies they had used to produce a square out of two, three or four triangles. Then the teacher asked the children to return to their tables, with a piece of paper, and draw a *Sketchpad* triangle. At first, a few children were unsure how to proceed, but after looking at what their classmates were doing, they set to work. Note that the task of producing a static drawing of a mobile situation can be quite challenging. Indeed, one of the goals of this task was to see how the children would communicate the temporal and dynamic nature of the *Sketchpad* triangle.

Children's static drawings of dynamic triangles

Figures 7.3 and 7.4 show a sample of the drawings that were produced by the students. It is striking to note the variety of techniques that the children came up with in their drawings. In general, there were two ways in which students approached the task of drawing: (1) drawings of multiple triangles, each one different from the other (Figure 7.3a–c); and (2) drawings of single triangles undergoing transformation (Figure 7.4a–c, on the next page).

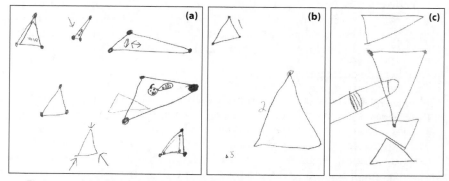

Figure 7.3: (a) Multiple triangles that can move in the double-arrowed directions; (b) Numbered triangles that evoke a serial change in shape; (c) The finger's role in moving the triangle.

Even within the multiple triangle approach, the students' drawings are unique. In Figure 7.3a, different types of arrows are used to show the motion of the triangle. The two arrows in the top right triangle communicate perpendicular directions of motions while the three arrows on the bottom left triangle indicate the mobility of the vertices of the triangle. Note that there are different types of triangle, including long and skinny ones.

In Figure 7.3b, there is a relationship between the triangles expressed in the numbering from one to three – with the triangle changing size in each transition. This numbering expressed a temporality and, more than Figure 7.3a, the sense in which one triangle might be undergoing changes. The finger in Figure 7.3c brings the subjectivity of the child into play, as she drags the triangle on the screen. There are multiple triangles shown, but much closer together than in Figures 7.3a and 7.3b.

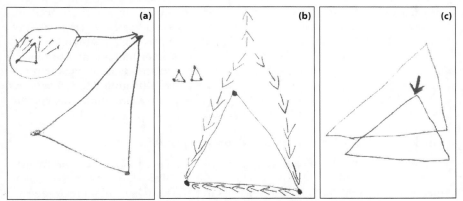

Figure 7.4: (a) Explosion at the vertex; (b) Motion in dotted-line arrows; (c) Motion through shading.

In Figure 7.4, the drawings are less about the discrete example space of the dynamic triangle and more about its continuous changeability. In Figure 7.4a, a little triangle in the top left has arrows indicating its growth into the larger triangle on the right. The circling of the small triangle and the arrow to the top vertex suggests a change in state from smaller to bigger – but where the smaller one no longer really appears on the screen, which would indeed be the case under dragging. In 7.4b, the arrows seem to suggest a potentiality of motion for all the sides of the triangle, with a new state hinted at by the outline of the arrows. Then, in 7.4c, the arrow and the fading of the top triangle also suggest a temporal change, with the faded triangle no longer visible, having moved into its new shape and location.

These drawings help us see how the children perceive the dynamic shape. The first category of drawings focuses more on the multiple different shapes while the second focuses on the continuous transformation between shapes. In terms of the children's spatial reasoning, we see that they are able to produce a range of triangles – and not just the prototypical one. The drawings in both categories show evidence of intrinsic-dynamic spatial reasoning as the children transform the triangle in time. They communicate this temporality with a variety of strategies including arrows, sequencing, and shading – all strategies that they used spontaneously, without formal instruction. These drawings not only help the teacher and researcher assess the children's spatial reasoning, but they provide opportunities for the children to develop a tangible discourse about triangles that captures the complex temporality of the dynamic diagram while also creating static artifacts that can in turn evoke new potential motions. These will be useful as the children begin to consider the conditions under which a triangle can become isosceles or equilateral, or even collapse to a segment.

Working with dynamic diagrams can be a fruitful way to help children attend both to the relationships that might exist between different mathematical objects as well as to the invariances that pertain in certain configurations. For example, Sinclair and Moss (2012) showed that by dragging a triangle constructed in a dynamic geometry environment, young children begin to attend to the fact that there are always three sides and three vertices no matter what the triangle's size, shape, and orientation. A frequent component of task design making use of digital tools involves having learners produce drawings that show some aspects of their interactions with the tools. From the point of view of the theory of semiotic mediation, this contributes to the process of meaning making, with a particular emphasis on the visual and dynamic aspects of the concepts under investigation.

Children's choice of representational tools: paper versus iPad

In the third vignette we continue our exploration of motion and markings as students in a Grade 3/4 class engage in a large-scale mapping project. Over the course of eight school days, students in Ms. Cordy's combined Grade 3/4 class completed a unit of study related to transformational geometry. The students engaged in a variety of learning activities including mapping tasks, dynamic geometry tasks on their iPads, and teacher-led inquiry-based tasks. The culminating summative assessment for the unit was an activity related to mapping.

Students were asked to create maps of their school that contained elements of transformational geometry such as objects rotating, reflecting, translating, and/or dilating. The map was to be used as a guide to locate the school mascot, Carlton the Colt, who was hidden somewhere in the school. Students had the option of creating the maps on large chart paper (84 cm by 63 cm) or on iPads. For assessment purposes, students were asked to describe their maps and their descriptions were captured on their iPads using an application that captured both sound and images.

Despite the fact that iPads were used daily in their learning, most students chose to create their maps on chart paper. The thirteen students who created their maps on chart paper and the six who did their drawing on the iPad all created proportionally similar numbers of objects in the maps. Students who produced chart paper maps had proportionally more verbalizations compared to those who produced iPad maps.

Egon's and Janine's mapping projects

In this unit, two provincially mandated curriculum expectations were explored: (1) understanding key concepts in transformational geometry using concrete materials and drawings, and (2) using language to effectively describe geometric concepts, reasoning, and investigations (OMET, 2005, pp. 45–46).

Egon (8 years old) created his map on his iPad (see Figure 7.5). He produced a total of 56 objects in his map, with nine of the objects depicting some aspect of transformational geometry. He also had nine dynamic verbalizations as he described his map:

> So, to find Carlton the Colt, you have to go west then go north all the way until you see the elevator. Eh, wait until it opens and he'll be here. Then he'll run out and go up the stairs. So follow him up the stairs. Yeah, and you'll be right here. Then, again, you go west, then you go north all past all the classrooms until you see the music room. You go in there, check behind all the desks there, and maybe a bit more, and you'll find him. So, check the door and trap him, so then you can grab him.

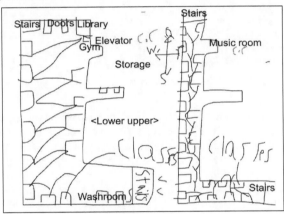

Figure 7.5: Egon's iPad map.

Egon accurately used cardinal directions and illustrates those on his drawing. The types of dynamic objects he includes are directional zig-zags to show the path toward Carlton the Colt, and identification of the stairs and elevator mentioned in his description.

Janine (8 years old) chose to create her motion map on chart paper (see Figure 7.6). Janine produced more objects in her drawing (62 objects in all) than Egon; however, only one of these objects showed evidence of transformational geometry. Overall, Janine had more to say about her drawing than Egon:

> Carlton the Colt is hiding in three places. Let's go find him. Okay, first, you're gonna go through the door. And then you're gonna, um, go this way. And, um, you're gonna go forward and, you will see that there are classrooms on each side of you, and you will see that there is the staff room. There's a place Carlton is hiding. Oh no! He ran away. Let's go search another spot. Okay. Now, let's go forward and, you're – there's the gym on that side of you and the library

on that side of you. Let's go to the library where Carlton is hiding. Oh! He got away again. Well, now we have to go out the door, so let's go past the staff room, and there's the door. So, we're gonna go out of the door and now we're gonna go in the portapack, and there's Miss Cordy's class. That's where Carlton is hiding. Oh! We caught Carlton!

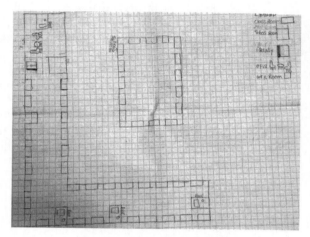

Figure 7.6: Janine's paper map.

Janine included a legend identifying the rooms. Her description was vague in terms of the cardinal directions of her path and even the teacher had difficulty following her path despite knowing the school. Her map primarily included hallways and classrooms and did not include paths as was seen in Egon's map.

Discussion of mapping activities

This vignette is drawn from a larger project that explored students' representation of motion in drawings (Kotsopoulos et al., 2014; Kotsopoulos, Cordy, & Langemeyer, under review). This large-scale mapping task served as an inquiry-type summative assessment in the class. Large-scale tasks (i.e., real-world environmental tasks) involve the person and are set in scenarios where the individual "cannot see the whole space of interest at once" (Quaiser-Pohl, Lehmann, & Eid, 2004, p. 104). In contrast, small-scale spatial tasks involve spaces whereby the person's orientation and movement is irrelevant to conceptualizing the space.

As part of the assessment, each child's drawing and verbal descriptions were assessed. The choice of drawing the map on an iPad versus on paper did not appear to have an impact on achievement level. Half of the students who created iPad maps and approximately 60% who created paper maps were assessed as high performers. The large-scale aspect of the task may have influenced students' preference for chart paper to create their maps because the chart paper appeared to provide more overall surface area for the drawings. In reality, this is not the case given that the iPad could have provided comparable space if the iPad drawing space were perceived by students as scalable.

Not surprisingly, scale is particularly challenging for young children (Vasilyeva & Huttenlocher, 2004; Uttal et al., 2013) and goes through considerable developmental changes with age (Frick & Newcombe, 2012). Previous research suggests that scaling in young children is a process of shrinking or expanding spatial relations (Vasilyeva & Huttenlocher, 2004). Recent research has shown young children may use a series of mental transformations when engaging in scaling tasks, and while accuracy increases with age, the use of mental transformations appears to be constant across development (Möhring, Newcombe, & Frick, 2014).

The participating students demonstrated a range in performance with regard to the representation of space. Students with limited performance often used small-scale maps (i.e., one hallway connecting to the next) to conceptualize the mapping project instead of the available larger space (Kotsopoulos et al., under review). For example, these students focused their search in one hallway rather than utilizing the whole school space. We speculate that this may have been related to their environmental exposure and perceptions of access within the space and was not affected by the choice of representational tool. As other research has shown, broader environmental experiences also inform how students engage in large-scale mapping tasks (Piaget et al., 1960; Clearfield, 2004; Clements, 2004).

One area that may have been affected by the choice of mapping tool was with regard to locating true cardinal directions on their maps. While there were iPad and paper users who had difficulty locating direction, we noted that the difficulty seemed more pronounced for students who created iPad maps. One possible explanation for the difficulty may be related to the shift from portrait to landscape that occurs with a turn of the iPad.

This vignette offers practical implications for teachers located in (i) the task structure, and (ii) understanding the limitations particular learners may face. The choice of iPad versus paper as the drawing surface appeared inconsequential in overall task outcomes for students. More impactful was the students' perception of space and their ability to scale. Many students were able to think of the space as a large-scale space, rather than a series of small spaces. Teachers can support large-scale mapping tasks by making the interconnectedness of small spaces within the larger space more explicit in the task design, and by engaging students in a familiarization of the large-scale space, where possible. Further research is needed to determine if the choice of drawing surface and tool is indicative of preference or reflects other aspects of spatial perception.

Linking ideas

Our discussion through this chapter does not suggest that formal concepts of motion need to be introduced earlier, but rather that children appear to approach these tasks using their familiarity with drawing, and extending and integrating invented and conventional symbols to portray motion. As with

Chapter 6, we have tried to provide examples of spatially rich contexts in which young children can engage in mathematically sophisticated ideas. The mobile dimension of the spatial reasoning involved in producing drawings, as well as in inventing or using techniques to pictorially describe motion, not only helped to engage the children (their hands and their bodies) but also invited attention to significant mathematical ideas such as transforming, varying, mapping, modeling, path-finding, and scaling.

Related to Bruner's (1966) enacted (action-based), iconic (image-based), and symbolic (symbol-based) modes of representations, the children's drawings provided an important means of transition from experience in the world to symbolizing the world. As Châtelet writes, these drawings can be seen as gestures "caught mid-flight" from the moving hand. Having been caught, they can then perform as visual modes of communicating mathematical concepts. The teacher may not only develop an appreciation for the power of these drawings, but also learn to help children connect their newly emerging signs into more mathematically conventional ones.

In the first and third vignette, we note that the drawings involved an important shift in dimension from 3D to 2D – from the experiential world to the 2D surface of the paper. As the next chapter explores, these shifts in dimension are significant in mathematics, but not frequently developed. For us, it is interesting how the concept of motion, as far as it was communicated in the drawings, helped retain something of a temporal dimension. There are perhaps interesting analogues between moving from 3D to 2D spaces and moving from the temporal to the static.

8

Interactions between three dimensions and two dimensions

KRISTA FRANCIS, WALTER WHITELEY

In brief ...

To effectively use spatial reasoning, children need to learn the conventions and interpretations of the 2D representations of the 3D objects and be able to move fluently between each dimension. 2D representations are commonly uncritically allowed to stand for the 3D object. This chapter explores the implications of this oversight.

Two- and three-dimensional reasoning

When Krista (author) was in a field course at the end of her third year of undergraduate studies in geology, she sat across from an outcrop of rocks created by a road-cut. Krista was well acquainted with 2D representations of geological formations and structures in textbooks. Yet when her structural geology professor sat beside her and asked what structures she saw in the outcrop, she was stumped. She was unable to recognize the 2D representations she learned in class in the actual situated rocks. The necessary transition from 2D representation to 3D was problematic.

Other chapters in this volume addressed reasoning and transformations that live within 3D space or within 2D space (or even one-dimensional space – the line). Here we pause to examine the conflation of 2D and 3D reasoning where the 2D representation has been uncritically allowed to stand for the 3D object: a critical connection that has been glanced over. It is common to encounter conventional 2D representations of 3D objects, in books, on computer screens, and sometimes even in studies of vision. Shared constructions of 3D objects from such 2D representations cannot be assumed, and ways that transformations of 3D objects for reasoning will appear in the 2D representations can be complicated – even inaccessible.

A common practice of assessing spatial reasoning is the use of 2D tests (see Chapter 2). The assumption that strength in 2D spatial reasoning translates to 3D underlies many of the tests of spatial reasoning – but is there any evidence for this? For example, consider a modification of the Thurston card rotation test described in Ekstrom et al. (1976) below which requires selecting the shapes that are a rotation of the image on the left. (See Chapter 6, for a discussion of mental rotation as the prototypical test of spatial reasoning.) The analogous comparison task for children would be assembling a jigsaw puzzle. The child picks up a piece, moves it close to a target match, rotates the piece – then tries a match by a further translation. It is not a pure rotation, and we are used to ignoring the center of rotation – just noticing the angle of rotation. Compare this with constructing the same transformation in a dynamic geometry program. A rotation must have a center, and it is difficult even for teachers to find the center for single rotation between the left figure and the next one – though it is obvious that angle of rotation is 90°. These tests assume that the person will automatically do added translations. There could be an added source of confusion if the person considers a planar "flip" – a rotation in 3D in many elementary classrooms.

Figure 8.1: Card Rotation Test.

The 2D card rotation test items and other 2D measures of rotation do not measure how well a child can rotate 3D objects in space. Compare the card rotation measure with building a Lego™ robot from an instructional booklet with 2D representations of a 3D L-shaped object (see Figure 8.2, on the next page). To attach the L-shaped object to the robot being built, the 3D object must be recognized from the instruction booklet's 2D representation of the desired object, and the actual object found in any orientation of the orange tray containing almost 100 different objects. The constructor must now orient the L-shaped object accurately in 3-space – and bring it into contact. By continually requiring moving between conventional 2D representations and reoriented views of 3D objects, the Lego task captures more of the complexity of the relationship between 2D and 3D representations. This is something that children of age 6 do, from Lego designs.

The purpose of this chapter is to examine how we reason about 3D objects supported by 2D representations – to call attention to the complexity and the learned conventions of that relationship and the spatial reasoning required for these tasks. First, we will discuss the ambiguity of 2D representations of 3D objects. Next, we will draw upon scientific examples from chemistry, biology, geology, and orthographic projections to illustrate the ubiquitous use of conventional 2D representations of 3D objects and the reasoning tasks they may support. We attend to the implied conventions and assumptions in the 2D representations, and the difficulties those can cause. In particular, cultural examples provide insight into how the conventions of 2D representation are

Figure 8.2: Attaching a 3D L-shape to a robot.

learned (or not). Next, we continue to challenge the assumptions that 2D is an adequate representation of the 3D world by examining children's transfer to a 3D world from 2D media like books and television.

Ambiguity of 2D representations of objects

Children are exposed to geometric shapes from early years. For example, teachers hang posters on the classroom walls showing plane and solid shapes. Solid shapes on the poster are representations of 3D shapes on a 2D surface – representations that we expect children to "see" the way we "see." Research on children's reasoning about mathematical diagrams suggests, however, that children's interpretations of such representations are not made in a straightforward manner (e.g., Steenpaß & Steinbring, 2013). In fact, a recent study showed that first graders do not necessarily interpret the 2D diagrams of 3D shapes in a standard, conventional way (Hallowell et al., in press). That is, the students had a difficult time "seeing" the components of a 3D shape from a 2D diagram. For example, when shown a 2D diagram of a solid pyramid, the majority of these students noticed the triangular shape but ignored the square on the bottom. Overall, the results showed that first graders had difficulty translating between 2D diagrams of solids and solid manipulatives.

Two key properties of 2D representations of 3D objects are that they are ambiguous and include learned conventions: we create what we see (Hoffman, 2000a). For instance, multiple 2D representations generate the same 3D object and multiple 3D objects can be represented by the same 2D image. Consider Hoffman's (2000a) example of the following three 2D "representations of a 3D cube."

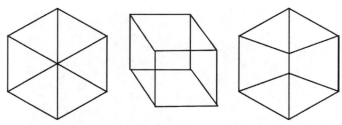

Figure 8.3: Hoffman's (2000a) example of 2D representations of a 3D cube (used with permission).

According to Hoffman most people see the middle picture as a wire-frame cube easier than they see the outer pictures as cubes because of visual rules or conventions. We usually observe 3D objects with stereo vision, or with a quick movement of the head or the eyes – separating incidentally overlapping points and lines. The representational convention is that overlaps in the 2D image are overlaps in 3D. Note that, geometrically, the middle image is not ever how the human eye would see a cube, with perspective, but we are conditioned to accept this. The learned rules dictate how we interpret depth of 2D representations. He argues that the interpretative rules are acquired by being exposed to visual experiences and other representations in the same way the rules of grammar are acquired by linguistic experience. Hoffman (2000a, 2000b) also argues that we have a preference for a simple 3D interpretation of a flat 2D image, over a more complicated collection of 2D objects in a shared plane.

A distinct ambiguity is that multiple 3D objects can be represented by the same 2D representation. Consider the possible 3D objects that could be attributed to the image on the right of the figure above. The hexagonal shape could be the view from the top of a gazebo-like shape. The middle figure is also ambiguous as a "necker cube" (Wikipedia, 2014c; turn your head or the page if you don't "see a second 3D shape," reversing which faces are at the front).

A further example is the "ambiguous box" below. The two parts of Figure 8.4 are "the same" (one is half-turned – turn the page or your head). Because we grew up with a "cue" that light comes from above (represented by the arrow), we initially "see" the lighter surfaces as "up." In A, this leads to a box in the corner of a room, the most common interpretation. We interpret B as a box with a cube cut out of a corner. Some people can also see a third possibility: a smaller box touching a bigger box along an edge – but the shading cues would now require two distinct light sources – one from the top and one from the bottom).

In practice, most of us use stereo vision and small movements of our heads, to disambiguate what we see into a 3D shape (in our brains). We are very good at this and do not think much about it, until we are faced with a truly flat image. These two examples highlight a fundamental classroom issue: what the teacher "sees" (interprets) may not be what many of the students interpret, and what one student "sees" is different from what another

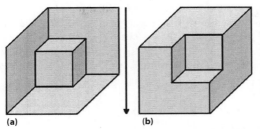

Figure 8.4: Two orientations of an "ambiguous box" (suggested light source above).

student "sees." In a classroom discussion, we need to attend to different possible interpretations that the students may be describing.

Visual perceptions vary across cultures. For instance, where a person was raised determines how you would answer the Müller-Lyer illusion below. Americans are more likely to find the bottom line with the arrows feathered out as the longest. Whereas people from the Kalahari are more likely to see the lines as equal length (Watters, 2013) for reasons drawing from distinct cultural experiences.

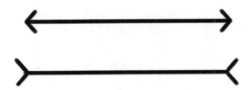

Figure 8.5: Which line is longer?[1]

2D representations of 3D objects in science

The correlation between spatial reasoning and success in STEM (science, technology, engineering, and mathematics) is well documented (see Stumpf & Haldimann, 1997; Clements & Sarama, 2011; Benbow, 2012; Sorby, 2012). This is reinforced by the use of spatial reasoning tasks in admissions tests and screening tests for STEM professions.

Children learn spatial reasoning – and cues – from 3D experiences, from 2D representations and their associated conventions (see Chapter 7), and from feedback during interactions with peers and adults. Fostering and developing these emergent skills in the early years is important for developing the spatial reasoning necessary for understanding advanced scientific and mathematical concepts, and problem solving (reasoning) with these representations of the concepts. The complex spatial reasoning skills that are necessary in many undergraduate STEM disciplines require the simultaneous use of many previous skills developed in prior years, beginning with skills developed and supported in the early years. If students have not developed sufficient spatial abilities, including moving back and forth among multiple presentations in

2D, with different conventions, in their problem solving – then they face an added barrier to success in these subjects. This is one reason why so many students struggle with these subjects. This next section provides examples of 2D representations of 3D information and reasoning tasks from chemistry, biology, and geology to highlight some of the complex spatial information presented in the 2D representations.

Chemistry: experiences with molecules

Although they had 2D diagrams of "bonded atoms" as early as the 17th century, in 1864 models of methane (CH_4) were still flat. As a field, it was only around 1874 that chemistry started working with 3D structures and associated representations (Figure 8.6). Today, fluency with stereo chemistry requires being able to convert 2D images of atoms and molecules into 3D objects for reasoning (stereo means 3D). The 2D representations of molecules include many assumed and learned conventions. Consider the following illustrations of the molecule ethane C_2H_6. The diagram represents sophisticated knowledge of electrochemical bonds.

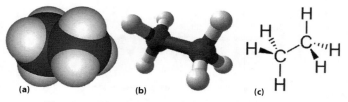

(a)   (b)   (c)

Figure 8.6: Three 2D representations[2] of ethane C_2H_6.

In Figures 8.6a and 8.6b, dark spheres represent carbon atoms and white spheres represent hydrogen. The solid lines joining the hydrogen and carbon represent single bonds (the sharing of one electron) (Stull et al., 2012; Stull et al., 2013). In Figure 8.6c, a solid triangle represents a bond towards the front, while a shaded triangle represents the bond is pointed back. (See also the triangles in Figure 8.7.)

As the molecules become more complicated, more notations and conventions of representations appear. Two representations of the molecule atisane appear in Figure 8.7. Both 2D illustrations represent the same 3D molecule. The image on the right is a slightly rotated version of the left. This time the bond going back is a dashed box, and the partial square at the top represents a bond aligned with the plane (Figure 8.7 – the Fischer convention from 1891; Stull et al., 2012; Stull et al., 2013).[3] These 2D representations of the molecules, ethane and atisane, exemplify the complicated and nuanced conventions of chemistry that must be learned in order to reason with its molecular structure (and to pass the MCAT exam for medical school).

Reflection in molecular 3D structure dictates very different chemical properties and behaviors. *Chirality* (i.e., handedness) is used to describe molecules that are distinct in 3D but are mirror reflections (Figure 8.8; Wikipedia,

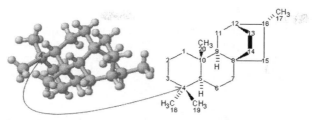

Figure 8.7: Two 2D representations of atisane (from slightly different perspectives).[4]

2014b). Spearmint and caraway have the same composition and similar bond structure. However, as exact mirror images, they taste and smell different – at the protein levels our bodies are chiral – without mirror symmetries. Chirality in drugs can be devastating. For instance, thalidomide is a sedative with one chirality and causes fetal abnormalities with the other chirality (Wikipedia, 2014b). Standard tests of 3D spatial rotation depend on the participants calling an object and its mirror image "different." However, notice that some molecules are *achiral* – the mirror image can be rotated to the original molecule, as happens in ethane (Figure 8.6). Strikingly, chirality of molecules does not appear in many high school chemistry curricula, or in mathematics curricula, although it is an important application of spatial reasoning in the world.

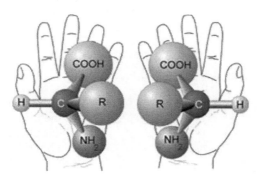

Figure 8.8: A chiral (handed) molecule and its mirror image are distinct
(under rotations) (NASA, 2008).

Pre-school children actively recognize this "difference" between mirror images in mental rotation activities. Our observations of 3-year-old children show that they know they cannot (rotate) a left shoe to a right shoe when putting them on, but they can flip (rotate in 3D) a cut-out of a right foot to a cut-out of a left foot. In Chapter 6, a vignette describes developing children's spatial reasoning skills by finding reflections/mirror shapes with dynamic geometry software. Consider the usefulness of these skills in order to later understand chiral molecules. Similarly, the Chapter 6 Cube Challenge lesson about constructing unique 3D shapes with multilink cubes has spatial skills used to understand how molecules are built and interact. Yet, it is too simplistic to consider that spatial reasoning skills are always straightforward to apply. Rather they are complex and emergent, requiring the simultaneous use of many previously learned skills.

Without strong spatial reasoning skills, reasoning with these diagrams and the molecular structures would be problematic. Developing strong spatial skills over multiple years can help students access and understand the notations and conventions of 2D molecular representations. Conflating flip (3D rotation) and plane reflection actually can become a barrier to critically important reasoning about handedness (chirality) in stereochemistry (stereo means 3D). In turn, as noted above, chirality and mirror images are a central issue in biology (our proteins are chiral), in drug design and in more general spatial reasoning (Wikipedia, 2014b). Reasoning with 2D representations of molecules is crucial to problem solving in chemistry.

Biology: exploring the inner ear

Like chemistry, biology also requires fluency with 2D and 3D conventions for understanding the 3D objects and interactions. An example of such conventions can be found in diagrams of the inner ear. These 3D biological structures also point out how evolution has developed a biological way to combine 2D embodied sensors for our 3D reasoning.

The inner ear is not visible, so most people's experiences of actually seeing the inner ear are limited or non-existent. The two images below are 2D representations of the right inner ear from above and from in front (anterior) (Wikipedia, 2014a).

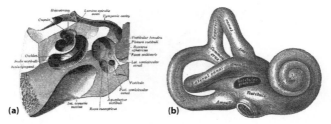

Figure 8.9: (a) Inner ear view from above; (b) Inner ear view from the front (Gray, 1858).

Without cues it might be difficult to see that the two diagrams above are two 2D representations of the same right ear. They look very different and use different conventions. The labeling cues help orient and the shared nomenclature supports developing a transformation of the shared 3D object within the two images. The cochlea and vestibule are labeled in each image. Notice that the cochlea is on the left in the first image and on the right in the second image, yet both images represent the right ear. The difference is the orientation of the 3D inner ear being represented and the correspondence requires being able to rotate the image mentally. Other available images show the location of the inner ear within the head and cross-sections of the inner ear. Compiling a whole picture from a cross-section, even with added context behind, requires mental assembly. The cross-section in Figure 8.9a uses shades of gray to distinguish different types of tissue. A great deal of learning of conventions is necessary for understanding anatomical diagrams.

A second look at the inner ear above finds that there are three planar canals (essentially aligned as semi-circles as three perpendicular coordinate planes). These canals give the brain three inputs of 2D angular acceleration to generate unambiguous information to support real-time cognitive conclusions of how the head is moving. These are deeply analogous to using 3D coordinate systems to represent and reason about 3D. This is one way human biology has evolved to capture 3D information with multiple 2D representations.

More generally, a lot of medical imaging uses computational geometry to reconstruct 3D objects and 3D connections from multiple 2D scans. Other medical imaging and robotic vision uses computational geometry to reconstruct 3D objects from several 2D projections.

Geology and geometry: 2D representations of 3D objects

Geology also requires considerable spatial reasoning for understanding 2D representations of 3D. The following exercise requires selecting the cross-sectional shape produced by a plane cut through a crystalline structure, represented in a 2D picture. The same type of representation occurs in the geometric study of polyhedra (which occur in both geology and crystallography – but far more broadly to support many forms of mathematical reasoning about patterns).

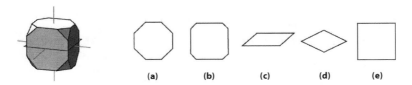

        (a)   (b)   (c)   (d)   (e)

Figure 8.10: Cross-section of a crystalline structure (Ormand et al., 2013).[5]

The crystalline example in Figure 8.10 is similar to the 3D cross-sectioning sample test item discussed in Chapter 3. Consider a few of the conventions that must be understood to visualize the 3D object from the 2D representation.

- Three lines correspond to the perpendicular horizontal (x and y) and vertical axes (z) of Cartesian space.
- The back is assumed to be a rotation of the front. Conventions of perspective imply the shown irregular 8-sided polygons are all congruent octagons with symmetries around the axes and the irregular triangles are congruent equilateral triangles.
- Shades of gray convey shading or orientation cues and imply a light source in the upper unseen left, in front of the plane of the vertical and horizontal axes.

This generates the square as the desired response – provided the test taker also orients the slicing plane parallel to the page. Without adding this convention of which plane the 2D representation is in, the parallelogram of item C is the more obvious choice.

Another common exercise in geology is selecting the appropriate configuration of layers of rock type within a rectangular cross-sectional shape in geologic block diagrams (Figure 8.11).

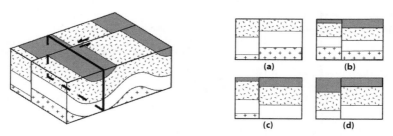

Figure 8.11: Cross-sections of a geologic block diagram (Ormand et al., 2014).

Selecting the appropriate answer in the block diagram above theoretically requires imagining a slice through layers of rocks in the earth. The lack of geologic context of scale, age, rock type, and possible origins of formation make the question somewhat ambiguous. Geometrically, an answer requires further 3D conjectures (guesses) about the interior of the block from a single 2D image, developed from three other plane sections (on three perpendicular faces). There are objects fitting the picture with cross-sections C and D. The arrows indicate a further convention on the slipping of the two sides of the fault line – suggesting C is the expected answer.

Both geological examples are problematic because of the underlying assumption that a student will be good at mentally manipulating 3D objects represented in a single 2D image, with multiple conventions. They might be good at appropriate 3D spatial reasoning with a spatial model. As described in the opening paragraph, Krista was quite familiar with (and good at) the types of cross-sectional diagrams in the test items above in her undergraduate geology degree. Yet she was not able to transfer such 2D representations and associated reasoning into understanding the actual structures of 3D outcrops. Spatial reasoning about geological rock formations occurs in 3D and in context, and can be difficult to grasp from 2D images in textbooks.

Mathematics and engineering: experiences with orthographic projections

Conventions in mathematics are learned, but often assumed. Orthographic projections are representations of objects parallel to one of the coordinate axes of the objects. Such projections are common in technical and architectural drawings. Multiple orthographic views (commonly three) are frequently used to disambiguate the projections (see Wikipedia, 2014a).

The conventions of engineering drawings use analogous multiple projections into three perpendicular planes. These conventions have been embedded in descriptive geometry for more than 200 years, since French military engineers learned to accurately represent enemy forts, so people at a distant location could plan their attacks. Though widely used, and sometimes taught in drafting classes in high school, reasoning with these multiple projections is assumed in tests of spatial reasoning for engineers, and careers with spatial-mechanical reasoning. These cultural practices embody both an essential cognitive fit of these effective representations and the artificiality of such cultural conventions that need to be taught and learned.

Without knowledge of the conventions, more orthographic projections might not help with reconstruction of the 3D shape. For instance, revisiting Hoffman's 2D representations of a cube that were introduced at the start of the chapter (and re-presented below), what would a child see?

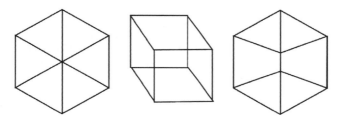

Figure 8.12: Hoffman's (2000a) example of 2D representations of a 3D cube.[6]

What strategies would be useful for helping them see a cube or hexagonal pyramid (on the right)? Working with actual 3D objects and the 2D representations could help them to learn conventions of representation. Regular transparent and hollow solids with varying levels of water have been used for observing various cross-sections of the solid (Mamolo, Sinclair, & Whiteley, 2011). Drawing the cross-sections within the 3D regular prisms could be another strategy for developing fluency moving between 2D and 3D space. Projecting shadows of 3D objects is another spatial exercise – all strategies initially supported by handling the 3D objects.

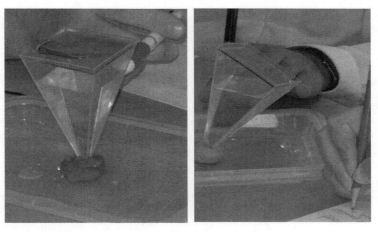

Figure 8.13: Learning about cross-sections through filling pyramids.[7]

In the next section, cultural examples from the Inuit and people from Papua New Guinea provide insight into how cultural experiences, different conventions and different languages contribute to 2D and 3D understandings.

Cultural experiences

Recent writings have pointed out that many psychological tests, including spatial tests, have been developed for people in Western, Educated, Industrial, Rich, and Democratic (WEIRD) cultures (Hendrich et al., 2010). Testing and teaching other cultures indicates important variations in contexts, language and significance of spatial reasoning – and therefore test performance.

Many cultures use patterns on surfaces (bead work, weaving, etc.) which have two-dimensional mirror symmetry, rotations and even local translations. However, in general these 2D experiences and vocabulary do not support 3D spatial reasoning, or transformations between 3D and its 2D representations. The world is 3D and 2D representations or even 2D objects can be artificially contrived components within 3D. For children who have not learned conventions in earlier years, understanding 2D representations can be difficult. Cultural experiences also influence the learning of conventions for 2D representations of 3D objects.

Inuit children's spatial reasoning

Poirier (2007) classified Inuit mathematical knowledge and developed curriculum for Inuit school children in Grades 1–3. She observed that Inuit children have strong spatial skills including an outstanding sense of space for locating themselves in the environments. They can read the snowbanks and assess the direction of winds. They can tell how far they are from the ocean/bay by smelling how salty the air is. The Inuit build *inukshuks* (human-shaped piles of rocks) that can be seen from afar as markers for orienting themselves in their environment and for messages about the landscape. These were all spatial reasoning, without substantial roles for 2D representations or 2D objects or language.

2D conventions of representation and spatial skills in Papua New Guinea

Bishop argues that the conventions of Industrialized (WEIRD) mathematics are not naturally acquired, but can be learned with training. Analyses of his Papua New Guinea (PNG) data indicated that often students who had grown up and had attended community schools in PNG villages were not aware of, and therefore had not learned, many standard mathematics conventions (Bishop as cited in Clements, 2008).

Bishop's (2008) study of spatial skills of students in PNG provides insight into the cultural nature of the conventions of 2D representations. Bishop studied twelve male first-year university students aged 16 to 26 in PNG to identify their strengths and weaknesses in spatial reasoning. He collected 6 to 7 hours of test data per individual. He found that there was unfamiliarity with the conventions of diagrams commonly used in Industrialized education:

> The representation of a three-dimensional object by means of a two-dimensional diagram demands considerable conventionalising which is by no means immediately recognisable by those from non-Western cultures. (p. 110)

Drawing tasks highlighted many of the Industrialized conventions that the PNG students were unaware of. When drawing a cube, several students had the correct square front, but the depth view was incorrect. Conventions for knowing how to represent depth were unknown. Another unknown convention was that dotted lines in a drawing of a cube can indicate edges at the back that are not visible. In another task when copying drawing from a specimen set, the scales and angles varied and the lines and curvatures were altered. While some of the variation could be attributed to unfamiliarity of drawing, the PNG students did not adhere to the Industrialized convention that "copy" means identical. Accuracy of replication had different standards.

Language and spatial skills in Papua New Guinea

Differences in language also likely contributed to the Papua New Guinea students' spatial understandings and difficulties with conventions. The students were asked to translate a list of 70 English words into their own local languages.

> [O]nly the following words were able to be translated by all twelve students: below, far, near, in front, behind, between, middle, last, deep, tall, long, short, inside, outside and hill. Some of the words that were omitted (i.e., difficult to translate, or forgotten) by more than half the students were: opposite, forwards, line, round, smooth, steep, surface, size, shape, picture, pattern, slope, direction, horizontal and vertical. (Bishop, 2008, p. 115)

Many of these words are frequently used for describing spatial and mathematical relations. Poirier's (2007) and Bishop's (1988b, 2008) research provide insight into how cultural, social, and language factors can have a profound impact on knowing conventions for representing 3D objects in 2D. This is an important message for teachers who are trying to develop spatial reasoning among their students – they can bring significant differences into a WEIRD classroom.

Spatial reasoning and digital media: the video deficit effect

A common assumption is that children are able to transfer 2D information from 2D media into the real 3D world. For instance, will a child recognize a koala bear at a zoo after seeing a picture of one in a picture book? Young children's transfer of learning between 2D and 3D contexts is a cognitively complex task with changes gradually across early childhood (Barr, 2010). Books, television, touchscreens and computers are specific settings for examining the transfer of 2D representation to 3D objects. In a review of recent research, Barr (2010) summarizes that

> children can imitate actions presented on television using the corresponding real-world objects, but this same research also shows that children

learn less from television than they do from live demonstrations until they are at least 3 years old; termed the video deficit effect. (p. 128)

Video deficit effect refers to how infants are less able to transfer learning from television and still 2D images to real-life 3D situations than they are able to transfer learning from face-to-face interactions. The video deficit effect appears after 6 months of age, peaks at around 15 months of age and is still persistent at age 3. An example of the video deficit is observed in Troseth and DeLoach's study (as cited in Barr, 2010) when children viewed a toy being hidden in a room on a television. Prior to the hiding exercise, toddlers were provided with an extensive orientation to the room. During the hiding exercise an adult hid a toy in the room, while the child watched in an adjacent room on a television screen. The 2-year-olds were unable to locate the toy, but the 2.5-year-olds were successful.

Lowrie (2002) explored how 6-year-olds make sense of screen-based images on the computer. Children were asked to interpret 2D representations with static and dynamic programs and relate the images to 3D objects in the environment. Some children could not make the links between the 2D representations and 3D objects. They argue that children need experiences developing the transfer between 2D representations and 3D objects away from the computer before linking in an ICT environment.

This suggests that many children do not learn all the conventions and skills for interpreting 2D representations of 3D objects through media alone. They also need experiences in the 3D world and parental or educator guidance to learn the conventions and transitions. The touchscreen and dynamic software activities described in Chapter 6 are examples that can guide children to interpret 2D representations of 3D.

Linking ideas

This chapter draws attention to the fact that conventions of 2D representations of 3D objects are learned and not naturally acquired. Recognition of the 3D object from its 2D representation cannot be assumed, and spatial reasoning about the 3D objects requires an unambiguous reconstruction of the object (at least mentally). Moving back and forth between 2D representations and 3D objects requires acculturation to conventions used, and developed practice of when and how to move between 3D and 2D. Fluency with moving between 2D and 3D space is essential for reasoning and connecting representations of scientific concepts. It is important that we do not simply move to a flattened version of 2D in textbooks and software assuming this is sufficient for developing fluency with 3D reasoning. It is also important that we engage with the reality that children have been learning to reason in 3D, often without 2D representations, and make connections to their current tasks that will draw on what they learned. Children start with 3D reasoning – and will need it for decades afterwards. The curriculum should support development and retention of 3D reasoning across the subjects and across the years.

In the topology of Chapter 2, there is no distinction between 2D and 3D.

Nor is there a particular place for the reasoning moves between 3D and 2D representations. This is something that should be developed further, both for pairs of static representations, and for dynamic reasoning. These moves are hard for humans – and also hard within computational geometry, since computers do not work well with ambiguity. Computer programming forces the user to make choices that eliminate the ambiguity, though computers can move easily from a 3D representation to selected 2D representations – something we rely on.

Chapter 3 emphasized the learnability of spatial reasoning. This chapter focused on the critical need to develop specific learnable conventions and fluid movement between 3D objects and their 2D representations. These learnable skills include many distributed forms of spatial sense-making used by middle school students in the context of engineering design and construction (Ramey & Uttal, 2014).

Chapter 4 discussed the historical roots for the schism found between elementary and secondary curriculum. Examples from this chapter from biology and chemistry briefly describe the strong spatial orientation required for interpreting 2D and 3D representations. The specific conventions of biology and chemistry (and other sciences) draw upon earlier foundational conventions from Euclidian and Cartesian notions, and their conventional representations. Working to ensure that children have fluency with the conventions before entering the secondary school system could help overcome the encountered schism.

As we move beyond English language schooling, this chapter noted that children from non-Industrialized communities will need explicit instruction to help them acclimatize to conventions for 2D representation in a WEIRD curriculum. Learning how people from other cultures move into industrialized conventions of 2D representations can help us learn more about how to support the development of 3D reasoning.

These reflections also indicate strong reasons to develop an early years curriculum that supports 3D reasoning, and develops the culturally required 2D representations in thoughtful ways. Coxeter et al. (1967) offer one such curriculum, drafted by experts with extensive experience with spatial reasoning over a variety of subjects and levels. It does start with 3D as the core, and develops 2D later, as it supports and evolves from the prior 3D experiences. A similar approach is embedded in the earlier and original Froebel Kindergarten curriculum, interestingly developed by an early childhood educator coming from crystallography and practiced in spatial reasoning (Brosterman, 1997).

Though software offers some advantages that may help ease children's transition from 2D representations to 3D reasoning, some of the use of technology requires critical pedagogical decision making about when and how to use it. For instance, the lesson in Chapter 6 used a touchscreen for manipulating 2D representations of 3D objects. Software with representations of the multilink cubes is used in another lesson described in the same chapter: building unique 3D shapes. The 2D representation of 3D objects that the children had prior experiences with is a useful tool for developing more fluid 3D reasoning, supported by many representational conventions. For example, one limitation is that the software does not have a consistently located light source, and the manipulations within the software may not attend to the light source – the shading is attached to the object, not a fixed light source. Software can offer many opportunities for screen viewing objects with added 3D information: correcting the shading cues, making parts transparent, converting from solid to wire-frame, and moving cross-section views. All of these, and more, are used in protein modeling software (Jmol). The software we use in education may help with learning some conventions, but not all without a larger pedagogical context.

Chapter 7 describes how drawing is useful for developing children's understandings of spatial concepts and relations to the 3D world. Children's own drawings become 2D representations of their 3D world. Drawing exercises are one way to learn conventions for 2D representations and reinforce development of 3D reasoning. Some of this is embedded in elementary curricula, but perhaps not taught with a strong context and/or linked to larger spatial reasoning goals, across the curriculum.

Chapter 9 will return to a number of these themes – including the critical need to support teachers as they develop facility with these transfers between dimensions to support their own spatial sense-making, and their students' spatial sense-making.

Notes

1 Wikipedia Creative Commons License: http://commons.wikimedia.org/wiki/File:M%C3%BCller-Lyer_illusion.svg

2 Wikipedia Creative Commons License.

3 We are grateful to Sophia Poscente for helping one of the authors understand the 2D representations of the molecule atisane.

4 Wikipedia Creative Commons License.

5 Used with permission.

6 Used with permission.

7 Permission granted by photographer Stewart Craven.

Section 4

And so?
What kind of research agenda might we need to pursue?

SECTION COORDINATOR: BRENT DAVIS

Chapter 9: Spatializing school mathematics

We close our discussion by revisiting and revising our original characterization of "spatial reasoning." We first offer a more nuanced description of the phenomenon that is especially appropriate for learning environments. We then use that description to speak to the spatialization of school mathematics. Finally, we point to some of the theoretical and empirical gaps that we perceive.

9

Spatializing school mathematics

BRENT DAVIS, YUKARI OKAMOTO, WALTER WHITELEY

In brief …

We close our discussion by revisiting and revising our original characterization of "spatial reasoning." We first offer a more nuanced description of the phenomenon that is especially appropriate for learning environments. We then use that description to speak to the spatialization of school mathematics. Finally, we point to some of the theoretical and empirical gaps that we perceive.

Whenever we gather, a central activity of our Spatial Reasoning Study Group is to review new and old video recordings of young children engaging in a range of tasks. In spite of our having watched, described, and interpreted so many of these episodes, a regular – and usually collective – response is surprise, even astonishment. Children's capacities to reason spatially can be amazing as they invent new, extend old, and blend established competencies to meet new challenges. These capacities offer extended opportunities to engage children, to develop spatial reasoning, and to connect and support a wide range of learning across mathematics, the STEM disciplines, and beyond.

As we have argued throughout the book, we believe that school mathematics should be structured to draw on and to advance such competencies. To those ends, we offer a few recommendations and provocations in this closing chapter. It is organized into three sections. We begin by revisiting a question raised in our opening pages, on the nature of spatial reasoning. In that discussion we present some of our current thoughts on how we might move beyond listings of aspects toward representing the emergent complexity of the phenomenon. With that backdrop, in the subsequent section we offer some commentary on what it might mean to "spatialize" school mathematics. For us, such a project of spatialization entails a mathematics curriculum that is better fitted to contemporary problem-solving and decision-making demands, that takes advantage of current tools and interfaces, and that is informed by emergent insights into human learning. Finally, we close with some commentary on a theory-and-research agenda.

The emergent complexity of spatial reasoning

In Chapter 1, we offered a preliminary list of active and intertwining processes that we see as characterizing spatial reasoning. That list was assembled a few years ago in our very first meeting, and it comprises locating, orienting, decomposing/recomposing, shifting dimensions, balancing, diagramming, symmetrizing, navigating, transforming, comparing, scaling, feeling, and visualizing.

This strategy of characterizing the whole of spatial reasoning by cataloguing its parts is, of course, well entrenched. It is most obviously manifest in the many tests and observation protocols that have been designed to study the phenomenon. However, the more we grapple with spatial reasoning, the more dissatisfied we are with the tactic of compiling lists of descriptors to capture its character. This point hits home for us every time we meet and whenever we present some variation of the preceding list to colleagues or students. Follow-up discussions inevitably veer toward "What about ...?" as one unmentioned aspect or another comes to mind. We have thus been exploring alternate schemes to describe and represent spatial reasoning, with the hope of moving beyond the isolation of observable and measurable aspects of the complex phenomenon of spatial reasoning to a notion that better matches the cognitive activities of learners in authentic educational contexts.

We hasten to add that we in no way mean to problematize or reject characterizations based on categorization or methodologies rooted in measurement. On the contrary, as we have underscored throughout the book, such approaches afford valuable insights. Our point, rather, is that care must be taken not to conflate the phenomena we seek to understand with the qualities we are able to isolate and study.

Put in quite different terms, we regard spatial reasoning as an *emergent* phenomenon, in the sense developed within complexity science (cf., Davis et al., 2008; Davis, 2011). An emergent phenomenon is a clearly discernible whole that cannot be fully comprehended by reducing it to its components. Such forms arise in the entangled interactions of many aspects, agents, or subsystems – and, within those interactions, new and unpredictable possibilities can arise. Those possibilities, in turn, can affect and occasion the entire system's current and future properties and behaviors.

Embracing the notion of emergence does not rule out the strategy of attending to sub-components. On the contrary, it emphasizes the importance of paying attention to such elements. However, it also compels attention to their interactions – and, more specifically, to the possibilities that arise as those elements are extended and blended. By way of cogent illustration, we might point to the conceptual distance between the isolated skill of mental rotation and imagined superposition of two simple shapes that is portrayed in Figure 8.1 (p. 122) and the rather more sophisticated constellation of actions that is hinted in Figure 8.2 (p. 123). The child in the latter figure interprets an image and then discerns, grasps, orients, translates and attaches an L-shaped piece to a grander L-shaped object.

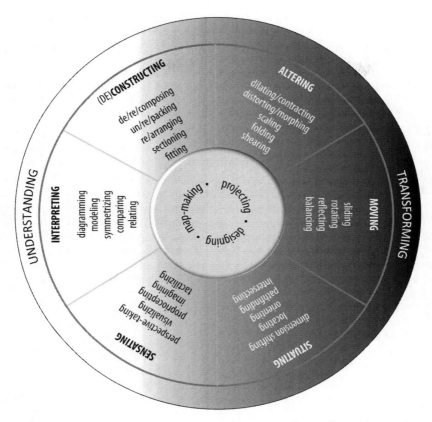

Figure 9.1: An attempt to represent the emergent complexity of spatial reasoning.

Figure 9.1 offers one representation of where we find ourselves at the moment in our efforts to move beyond lists of isolatable and testable aspects into a more emergent space. As with our earlier strategy of assembling a list of elements, there is no attempt to be exhaustive here. We are well aware that important elements of spatial reasoning have not been included in this graphic. At the same time, the image affords opportunity to gesture toward some critical qualities of spatial reasoning, including:

- the co-involved, complementary, and inextricable natures of "mental" UNDERSTANDING and "physical" TRANSFORMATIONS, signaled by their shared locations on the unbroken outer ring (with aspects more typically associated with understanding in the lighter regions, and aspects more typically associated with transformations in the darker regions);

- the entangled natures of the sub-elements of spatial reasoning (signaled by their juxtapositions – both within sub-lists and across those sub-lists – and by continuums of light and dark rather than black/white regions);

- the emergent-and-leveled nature of elements (signaled by boldfaced sub-headers, which we regard as more emergent capacities that arise in the blending and extending of the sorts of elements collected beneath those headers); and

- some possibilities for other levels of emergent competencies (signaled by the actions collected in the center) that arise in the blending and extending of boldfaced elements.

Not without irony, we note that this image makes sense only to the extent that observers are able to agree to particular conventions of spatial representation and interpretation. That is, the diagram is "spatial" (in the plane), but a number of the spatial properties in the image do not carry information for reasoning. For example, the three terms in the central circle could be rotated to any position; they are equally connected to all of the sectors of the next layer out. Similar could be said of the outside circle. Indeed, if the medium of the printed page supported the possibility, different aspects of the image would be merging, blending, and/or spinning. Once again we are reminded of the conceptual distances and spatial contrasts between the dynamic 3D world we inhabit and the constraints of the static 2D world we employ in print-based communications.

We opted for a circular layout in an attempt to avoid the suggestion of conceptual scaling, which is a typical implication of (i.e., a spatialized metaphor often associated with) a linear representation. Of course, a curved image does not entirely remove the possibility of such a reading. The risk of over-interpreting a spatial metaphor is still present in the circular adjacency – and even in the apparent tension between transforming and understanding.

We re-emphasize that this depiction of spatial reasoning is intended more to be provocative than comprehensive. We aim to provoke conversation among educators about how to move beyond the discretized characterization of spatial reasoning that is appropriate for the sanitized psychology laboratory setting into the messy and remarkable world of human problem-solving. Our students rapidly switch among various cognitive acts when drawing upon different forms of spatial data, just as they switch among representations when using spatial reasoning to solve problems. We believe there is potential to develop improved notions of spatial reasoning for pedagogical settings. For us, it is not an attempt to capture all its aspects, but a means to make sense of how spatial reasoning competencies arise, blend, and self-transform.

Consider, for example, what might be involved with the commonplace expectation of children to recognize the relationships between a curved surface and a flattened projection. A standard example is a spherical globe beside a 2D Mercator projection of the earth. Moving between such representations involves multiple and simultaneous TRANSFORMINGS (i.e., on the right side of Figure 9.1), including MOVINGS (e.g., ROTATIONS), ALTERINGS (e.g., DISTORTING and SCALING), and SITUATINGS (e.g., LOCATING and ORIENTING). At the same time, several elements of UNDERSTANDING (i.e., the left side of the figure) are invoked, such as INTERPRETING (e.g., COMPARING and RELATING) and SENSATING (e.g., VISUALIZING). Phrased differently, the act of TRANSFERRING and TRANSFORMING spatial information from one representation to another, while clearly an act of spatial reasoning, seems to operate on a different conceptual plane from, say, SLIDING (TRANSLATING).

Such examples abound as we move from the more parsed and controlled realm of the psychological test to the complex and situated world of children solving problems in the classroom. The model presented in Figure 9.1 is a first effort to capture that leap in complexity. We invite others to think with us on how such emergent complexities might be otherwise represented in manners that could be educationally productive.

Spatializing school mathematics

As we consider how spatial reasoning might be incorporated into school mathematics, we are aware of a continuum of possibilities. At one extreme, aspects of spatial reasoning might be selected and located within established programs of study in order to support the development of familiar topics. At this pole, for example, children might engage with multiple interpretations of "number" (e.g., as count, as measure, as position in space) through dynamic simulations that invite blended, diversified, and more robust understandings. In more general terms, elements of spatial reasoning such as those identified in Figure 9.1 might be taken up in one way or another to support the learning of virtually every concept in current standardized curricula.

As we develop our line of thought in this section, even though we would regard such acts of adding spatial representations and simple spatial reasoning as significant and valuable advances to the teaching and learning of mathematics, for us they represent a trivial embrace of spatial reasoning. We argue for a somewhat more radical take-up, one that more fully incorporates spatial reasoning as a central process that is worthy of including and developing to high levels for its own sake and as a contribution to integrated problem solving across mathematics and beyond. That is, we envision a major curriculum reconceptualization that might, among other ends,

- be better fitted to problem-solving and decision-making demands of daily life,

- take advantage of and contribute to current tools and interfaces that support and rely on spatial reasoning,

- enfold emergent insights into the roles of gesture, bodily aspect, and other elements of embodied cognition for mathematical understanding, and

- pull school mathematics into stronger alignment with practices in its parent disciplines in the mathematical sciences.

To our analysis, such goals entail a dramatic reformatting of school mathematics. That said, we acknowledge that we are not adequately positioned to propose a comprehensive program of studies that blends entrenched elementary foci of arithmetic and algebra with the complex competencies depicted in Figure 9.1. Other strands such as measurement and geometry should also be infused with spatial reasoning which is often lost when these areas are made servants of arithmetic and then algebra. Such a project would require the participations of categories of expertise that greatly surpass those represented in our small group.

Nevertheless, we do feel able to speak to some preliminary considerations – ones that are oriented by the stunning capacities of young learners. For example, children as young as age 3 have been observed to use geometric cues to locate objects on maps (Vasilyeva & Bowers, 2006). As discussed at the end of the preceding section, such competencies involve a complex blend of subskills, and each of those subskills is a sophisticated competency in itself. Map-reading, then might serve as an exemplar for how to approach the project of spatializing the curriculum. It presents opportunities to identify (and, subsequently, to develop and extend) subsets of skills while it presents occasions to blend those subskills into complex competencies.

Our strategy of interpreting elements of spatial reasoning as "layered" or "nested" is offered in contrast to typological frameworks used to describe the research findings, such as the one developed by Uttal et al. (2013), presented in Chapter 2 and applied in Chapters 2 and 3. One of the issues with that framework, as noted in those chapters, is that spatial skills in one category may shift to another category depending on how tasks are interpreted or performed. It is often cognitively adaptive to rapidly switch between mental activities associated with relevant cells, depending on the spatial task at hand. Keeping this in mind, and with a view toward a more radical spatialization of school mathematics, the following sort of activities may be appropriate for strengthening more than one category of spatial skills.

As discussed throughout the book, children in the first few years show remarkable developmental changes in their visual and spatial understanding. By age 4, most children begin to recognize and work with common geometric shapes such as a circle (Clements et al., 1999). Children's actions seem to be limited to visual characteristics of geometric shapes as opposed to reasoning. Thus, simply asking children to name familiar geometric shapes is not enough to promote spatial thinking. Children must also be alerted to *properties, transformations that preserve the essential features,* what varies and what is invariant. This is Felix Klein's now-classic definition of what makes a study geometry. The goal is that they understand, for example, what makes a triangle a triangle and not a rectangle. One potential activity includes asking children to sort plane geometric shapes, first by visual inspections, but then moving on to discussing what common characteristics are like. For example, improved understandings of the geometric properties of shapes might be fostered through a discussion about how two different sizes of equilateral triangles are similar to and different from each other, or how they differ from right triangles. Another valuable tool might be dynamic geometry explorations, where dragging reveals what changes and what is unchanged – provided, of course, the objects are carefully constructed (not drawn) to hold the desired features. Having a repository of such sketches, along with analysis for teachers would be a great resource.

Geometric shape puzzles can also be effective as long as the activity goes beyond a trial-and-error approach. Teachers and parents should encourage children to ask why a particular shape would fit in the space provided, for example, by bringing children's attention to the number of angles in the puzzle piece and the puzzle space. Similar approaches can be taken with familiar solid shapes as well. All of these activities could also strengthen children's mental rotation skills when comparisons between the initial shape and rotated shape are made. Providing a story (e.g., a square is the body of the house and a triangle is a roof) may prove effective (e.g., Casey, Andrews et al., 2008), as long as the story refers to geometric properties. Looking back, Froebel's activities with "gifts" of spatial collections built this from the initial programs for Kindergarten (literally, "children's garden").

As discussed in Chapter 2, research into mental rotation has resulted in some inconsistent findings in terms of developmental trajectory as well as gender differences. Although more research is clearly needed to reconcile contradictory findings, infants as young as 3 months old have been found to show sensitivity to object rotation (some of these early findings also reveal gender differences during infancy). If this is indeed the case, children, and girls in particular, may require ample experiences with dynamic rotations of objects. Training has been found to improve children's mental rotation skills (Uttal et al., 2013). Infant sign language and continuing use of sign language includes such practice of mental rotation, and is shown to produce stronger abilities. Thus early training in this area is important and likely to result in improved performance. Training could take advantage of existing measures of mental rotation for children.

Mental rotation activities could also be easily incorporated into a game context. For example, a teacher can present one simple shape using multilink cubes and show it to a class (e.g., a tromino in which two cubes on the bottom and another being on top of the bottom right cube). Then present the class with another tromino configured the same as the first but rotated 180°. The class can discuss if the second shape would look the same if rotated. During class discussion, it is important to describe the shape such as "on top of" or "to the right" as use of spatial language in early years is found to predict later spatial and mathematics performance (Pruden et al., 2011). Mental rotation activities could require seeing the same shape from another's perspective, which is characteristic of self-to-object navigation.

As we attempted to highlight in Chapters 6, 7, and 8, many other activities can be generated as one examines what each category of spatial skills entails. Activities that have already been part of typical instructional practices (e.g., patterning) can also be modified to emphasize the development of spatial skills. Educators need to pay close attention to opportunities that afford such experiences. However – and, in our view, more importantly – attention needs to extend well beyond entrenched curriculum foci. For example, as developed in Chapter 8, attention might be directed to the overlooked but critically important task of learning to use 2D representations to support 3D spatial reasoning. Because they push against convention, such hopes require design, research, and support.

And they should start early. In particular, schools and children need resources and teacher development that support this more radical sort of curriculum spatialization (recall Froebel). Materials and activities should support a sense of the importance of spatial reasoning, an awareness of its learnability in its various forms that we have outlined in Figure 9.1, and an appreciation of the roles of multiple representations (and multi-model representations and thinking) in children's development of spatial reasoning. Such goals entail major changes for teachers and classrooms, an expanded vision of what mathematics is about, and a curriculum that grounds and integrates these elements.

To re-emphasize, a critical aspect of what we propose is that spatial reasoning should be developed *for the sake of better spatial reasoning* – that is, not just in the service of better number sense or in anticipation of its utility for teaching algebra. For us, that is the essential difference between a trivial and a more radical spatialization of school mathematics. Of course, the point is *not* that spatial reasoning should be taken on as a discrete or self-contained topic. Quite the contrary, as we have argued throughout the book, it is a core, integrated element in mathematical thinking. As such, we in no way regard spatial reasoning as "one more" topic for school mathematics. Rather, we see it as a transformative force, one that should compel rethinkings of every aspect of contemporary school mathematics. Robust spatial reasoning skills in mathematics will also extend to better support mathematical reasoning in the sciences than the current discursive approach offers, affording an increasingly unified curriculum across domains. Our approach to thinking about spatial reasoning provides an opportunity to improve the pedagogical situation beyond mathematics.

At the same time, we do not wish to lose sight of emerging insights into the learning and teaching of spatial reasoning capacities. They can be deliberately developed. In particular, as we attempted to portray in Figure 9.1 by juxtaposing clusters of sub-competencies, progress is being made on identifying blends of spatial abilities – and, hence, designing experiences that support such blending. In pedagogical terms, we perceive that the mathematics education community is on the cusp of developing a collection of trajectories – that is, a curriculum – that will dramatically surpass what might be learned through everyday experiences and random practice.

Articulating a research program

That last assertion – that the mathematics education community may be converging on a powerful curriculum for spatial reasoning – also frames our thoughts on the sorts of theorizing and research that we perceive as necessary. We close the book by pointing to four topics that will be occupying much of our attention over the next several years – and that, we hope, might capture the attentions of other researchers and teachers.

1. Expanding the example space

First and foremost in our assessment of necessary developments to extend the field is the need for a broadened range of examples on how a radically spatialized mathematics curriculum might be lived out in classrooms. We have offered a handful of such cases in this book. Although few, we believe they actually represent a significant increase in the number that are readily available. To our reading, while there is a large number of rich examples in the current mathematics education literature, there is a relative lack of documented research into their impacts. We attribute this a dearth in large part to the fact that the topic of spatial reasoning tends to be engaged trivially, "in support of" entrenched curriculum emphases, rather than radically, "a challenge to" those emphases.

Notably, there is a parallel lack of nuanced examples of teachers' professional development. This absence is an interesting one, given that teachers' disciplinary knowledge of mathematics is arguably the most prominent and active area of inquiry among mathematics education researchers at the moment. The fact that there are only few illustrations of the obstacles to supporting teachers and future teachers in spatial reasoning (Whiteley et al., 2008) may be evidence of the overwhelming tendency to embrace spatial reasoning in a trivial way – that is, in support of rather than as a challenge to entrenched curricula.

We expect that elaboration of these "example spaces," as they emerge, will embody the realization that spatiality is not that restricted; it is present in virtually everything that humans are already doing. What is critical, then, are strategies for recognizing, representing, and triggering spatial reasoning, and then connecting these into networks of powerful reasoning. For instance, we are eager to encounter and examine lesson formats that render embedded and embodied spatiality available to conscious interrogation and deliberate elaboration.

Such an example space would serve a number of purposes. For instance, it would support the articulation of design principles to aid in structuring experiences and sequencing activities. Closely related, a robust and extensive example space would contribute to better understandings of paths of development, relative importance of subskills, and possibilities for emergent skills.

2. Studies of spatial reasoning in other settings

As hinted in Chapter 8, there is a certain temptation to assume that the WEIRD world is the most advanced across almost all domains, including (and, perhaps, particularly) perceptions of the surrounding world.

This is an assumption that has been challenged by anthropologists (e.g., Diamond, 2012), psychologists (e.g., Henrich et al., 2010), mathematicians (e.g., Eglash, 1999), and mathematics educators (e.g., D'Ambrosio, 1985). Of particular relevance, it seems that humans' perceptions are strongly influenced by the shapes of the spaces they inhabit. It is thus that residents of highly rectilinear Western societies are much more inclined to project lines and right angles onto the world, whereas people whose surroundings are more "natural" draw on very different forms to interpret and structure their worlds (see Chapter 8).

More locally, it will also be important to consider the role of spatial reasoning in domains beyond mathematics and other STEM fields. For example, Sorby (2009) has offered exercises for Engineering students that overlap with isometric projections that are in some elementary school curricula, as well as in the conventions of Lego instructions. There is increasing awareness that lack of spatial reasoning is a barrier to success in university studies, and that remedial work is needed when the curriculum has failed to develop and maintain spatial reasoning as a core competency.

On this matter, Black, Turner, and Bower (1979) offered evidence that spatial reasoning plays a key role in narrative comprehension and memory across every domain of human engagement. For example, in an experiment on reading comprehension, they asked people to read the following sentences:

- John was working in the front yard then he went inside.
- John was working in the front yard then he came inside.

The two sentences could be argued to mean exactly the same thing, but participants took longer to read the second sentence than the first. It was only some time afterward (Black et al., 2012) that the researchers linked that difference to the necessary spatial shift between the two sentences – and to the inherent conflict of "coming inside" from an exterior location.

The essential point hearkens back to the embedded and embodied natures of beings, which extends well beyond the realm of the mathematical. The need for a much elaborated research base is evident. However, once again, we have few ideas on where research into such matters may go. We have no doubt that there is much to be learned about aspects of spatial reasoning that might be underdeveloped, ignored, or simply unnoticed in the Western world.

3. More nuanced understanding of the interplay of numeracy and spatiality

Over the past several decades, great strides have been made toward better understandings of the bodily bases of mathematics and the conceptual tools that enable knowers to elaborate corporeal experience into sophisticated, abstract knowings. Spatial reasoning has proven to be a linchpin in these discussions (cf. Gunderson et al., 2012). For example, the capacity to conceive of "number" as a distance or as a location in space or a measure of the spatial concept of area is essential to moving from the most basic sort of arithmetic into more advanced topics such as pre-calculus and calculus. These topics such as optimization can also be addressed directly with spatial reasoning on volume and area without numbers (Mamolo & Whiteley, 2012; Whiteley & Mamolo, 2014).

Such insights have helped to interrupt centuries of habit of interpreting topics in elementary school arithmetic strictly in terms of operations on discrete quantities – wherein number is understood as count, addition as putting together, subtraction as taking away, multiplication as repeated putting together, and so on. Virtually all other interpretations of these operations have a spatial component (e.g., addition as directed movement; multiplication as area-making or scaling). Yet, not only are teachers largely unaware of the spatial dimensions of these concepts, they are typically unable to name alternative interpretations of concepts (Davis, 2011).

There is some irony here. As detailed in Chapters 5, 6, and 7, research into gesture reveals that educators and mathematicians frequently make use of not-necessarily-conscious actions to communicate and represent their understandings of mathematical concepts. For example, they often gesture without conscious effort to demonstrate magnitude and direction when describing mathematical relations. That is, simple observations provide evidence of a spatialized component of mathematical understanding that is much more present and much more pronounced than knowers are typically aware.

Geometry, and measurement, can be addressed with core spatial reasoning abilities and problem solving – but these are often shifted over to sources of algebraic reasoning rather than following through how spatial reasoning solves the problems and makes the answers "sensible" – accessible to the senses.

4. A detailed, networked analysis of research, with a consolidated vocabulary

Finally, we close by re-emphasizing the importance of broad and transdisciplinary research to extend insights into the nature of spatial reasoning and its role in the development of mathematical understanding. Interdisciplinary work is already happening, and in recent years we have witnessed (and participated) in meetings of experts in psychology, mathematics, and mathematics education. That said, other pockets of emergent expertise remain relatively isolated. For example, some especially powerful developments have been reported in science and engineering education at the tertiary level (Sorby, 2009, 2012; Sorby et al., 2013) – and as relevant as such developments are to the sort of work reported in this book, they can feel far removed.

One aspect of that experienced distance is a dissonant vocabulary. Briefly, research into spatial reasoning may be somewhat stymied by inconsistent vocabularies (both using different terms to refer to the same thing and the same term to refer to different things). In particular, as we consult literatures from other domains, we are keenly aware that such core terms as *spatiality* and *visualization* are used in very different ways. At the same time, we are confronted with completely unfamiliar phrases that, we suspect, are intended to signal some of the same phenomena that we believe we are observing. Perhaps a next important stage in the evolution of this research domain, then, is an enlarged conversation that juxtaposes interests, insights, and vocabularies from across fields and educational levels. The framework proposed by Uttal et al. (2013) is helpful in this regard, but as we point out above, it strikes us a bit myopic to be used in educational contexts.

To this end, among the upcoming projects we have identified for the Spatial Reasoning Study Group is a network analysis of research to date through a comprehensive review of the studies across domains and various interdisciplinary communities. These domains include Education, Psychology, Sociology, Anthropology, Neurology and Cognitive Science, Mathematics, Statistics Education, Visual Literacy, Arts Education, Computer Science, and other STEM disciplines (SIGGRAPH, 2002). The goal is to offer a coherent presentation of what has and has not been researched, how insights complement and collide, and how findings have been named and represented.

For us, these are exciting times, as we witness swelling interests in spatial reasoning across disciplines, observe its increasing relevance in the day-to-day world, and come to terms with its role in the teaching, learning, and doing of mathematics. Additionally, enormous advances in technology in the last decade open the door to new spatial visualization, in particular dynamic spatial presentation potentialities in the classroom (Sinclair, 2014). We look forward to the expanded conversation with great anticipation.

References

Alibali, M. W., & Nathan, M. J. (2007). Teachers' gestures as a means of scaffolding students' understanding: evidence from an early algebra lesson. In R. Goldman, R. Pea, B. Barron, & S. J. Derry (Eds.), *Video research in the learning sciences*. Mahwah, NJ: Erlbaum.

American Psychiatric Association. (2013). *Diagnostic and statistical manual of mental disorder (5th ed.)*. Washington, DC: American Psychiatric Publishing.

Atiyah, M. (2002). Mathematics in the 20th century. *London Mathematics Society, 34,* 1–15.

Baartmans, B. G., & Sorby, S. A. (1996). Making connections: spatial skills and engineering drawings. *The Mathematics Teacher, 89,* 348–357.

Baddeley, A. D. (1986). *Working memory*. Oxford, UK: Oxford University Press.

Baddeley, A. D., & Hitch, G. (1974). Working memory. In G. H. Bower (Ed.), *The psychology of learning and motivation: advances in research and theory* (Vol. 8, pp. 47–89). New York: Academic Press.

Baddeley, A. D., & Logie, R. H. (1999). Working memory: the multiple component model. In A. Miyake & P. Shah (Eds.), *Models of working memory: mechanisms of active maintenance and executive control* (pp. 28–61). New York: Cambridge University Press.

Bagust, J., Docherty, S., Haynes, W., Telford, R., & Isableu, B. (2013). Changes in Rod and Frame Test scores recorded in schoolchildren during development – a longitudinal study. *PLoS ONE, 8*(5): e65321. doi: 10.1371/journal.pone.0065321

Balacheff, N. (1988). Aspects of proof in pupils' practice of school mathematics (D. Pimm, Trans.). In D. Pimm (Ed.), *Mathematics, teachers and children* (pp. 216–235). London: Hodder & Stoughton

Baldy, R., Devichi, C., & Chatillon, J.-F. (2004). Developmental effects in 2D versus 3D versions in verticality and horizontality tasks. *Swiss Journal of Psychology, 63*(2), 75–83.

Barr, R. (2010). Transfer of learning between 2D and 3D sources during infancy: informing theory and practice. *Developmental Review, 30*(2), 128–154. doi: 10.1016/j.dr.2010.03.001

Bateson, G. (1972). *Steps to an ecology of mind: collected essays in anthropology, psychiatry, evolution, and epistemology*. Chicago: University of Chicago Press.

Benbow, C. P. (2012). Identifying and nurturing future innovators in science, technology, engineering, and mathematics: a review of findings from the study of mathematically precocious youth. *Peabody Journal of Education, 87*(1), 16–25. doi:10.1080/0161956X.2012.642236

Bénézet, L. P. (1935a, 1935b, 1936). The teaching of arithmetic I, II, III: the story of an experiment. *Journal of the National Education Association, 24*(8), 241–244; 24(9), 301–303; 25(1), 7–8.

Berger, S. E. (2010). Locomotor expertise predicts infants' perseverative errors. *Developmental Psychology, 46*(2), 326–336.

Bernstein, R. J. (1983). *Beyond objectivism and relativism: science, hermeneutics, and praxis*. Philadelphia, PA: University of Pennsylvania Press.

Bishop, A. J. (1980). Spatial abilities and mathematics education: a review. *Educational Studies in Mathematics, 11,* 257–269.

Bishop, A. J. (1983). Space and geometry. In R. Lesh & M. Landau (Eds.), *Acquisition of mathematics concepts and processes*. New York: Academic Press.

Bishop, A. J. (1986). What are some obstacles to learning geometry? *Studies in mathematics education,* UNESCO, 5, 141–159.

Bishop, A. J. (1988a). A review of research on visualization in mathematics education. In A. Borbás (Ed.), *Proceedings of the 12th Annual Meeting of the International Group for the Psychology of Mathematics Education, 1,* 170–176.

Bishop, A. J. (1988b). Mathematics education in its cultural context. *Educational Studies in Mathematics, 19*(2), 179–191.

Bishop, A. J. (2008). Visualising and mathematics in a pre-technological culture. In P. Clarkson & N. Presmeg (Eds.), *Critical issues in mathematics education* (pp. 109–119). New York, NY: Springer. Retrieved from http://link.springer.com.ezproxy.lib.ucalgary.ca/chapter/10.1007/978-0-387-09673-5_8.

Bishop, A. J., Clements, K., Keitel, C., Kilpatrick, J., & Laborde, C. (1996). *International handbook of mathematics education*. New York: Springer.

Black, J., Segal, A., Vitale, J., & Fadjo, C. (2012). Embodied cognition and learning environment design. In D. Jonassen & S. Land (Eds.), *Theoretical foundations of learning environments, 2nd ed.* (pp. 198–223). New York: Routledge.

Black, J., Turner, T. J., & Bower, G. H. (1979). Point of view in narrative comprehension, memory and production. *Journal of Verbal Learning and Verbal Behavior, 18,* 187–198.

Bomba, P. C. (1984). The development of orientation categories between 2 and 4 months of age. *Journal of Experimental Child Psychology, 37,* 607–636.

Booth, J. L., & Siegler, R. S. (2008). Numerical magnitude representations influence arithmetic learning. *Child Development, 79*(4), 1016–1031.

Boswell, S. L. (1976). Young children's processing of asymmetrical and symmetrical patterns. *Journal of Experimental Child Psychology, 22*(2), 309–318.

Bowers, C. A. (2011). *Perspectives on the ideas of Gregory Bateson, ecological intelligence, and educational reforms*. Eugene, OR: Eco-Justice Press LLC.

Brosterman, N. (1997). *Inventing Kindergarten*. Self published.

Bruce, C., Flynn, T., & Moss, J. (2013). A "no-ceiling" approach to young children's mathematics: preliminary results of an innovative professional learning program. In M. Martinez & A. Castro Superfine (Eds.), *Proceedings of the 35th annual meeting of the North American Chapter of the International Group for the Psychology of Mathematics Education*. Chicago, IL: University of Chicago.

Bruce, C. D., & Hawes, Z. (in press). The role of 2D and 3D mental rotation in mathematics for young children: What is it? Why is it important? And what can we do about it? *ZDM: The International Journal on Mathematics Education*. 10/2014; doi: 10.1007/s11858-01406374?sa_campaign=email/event/articleAuthor/onlineFirst

Bruce, C. D., Moss, J., Sinclair, N., Whiteley, W., Okamoto, Y., McGarvey, L., Drefs, M., Francis-Pocente, K., & Davis, B. (2013). *Early-years spatial reasoning: learning, teaching, and research implications*. Research symposium presented at Research Presession Conference of the Annual Meeting of the National Council of Teachers of Mathematics. Denver, CO.

Bruner, J. (1966) *Toward a theory of instruction*. Cambridge, MA: Harvard University Press.

Bryant, P. E. (2008). Paper 5: Understanding spaces and its representation in mathematics. In T. Nunez, P. Bryant, & A. Watson (Eds.), *Key understandings in mathematics learning: a report to the Nuffield Foundation*. Retrieved 28.04.2013 from http://www.nuffieldfoundation.org/sites/default/files/P5.pdf

Bryant, P. E. (1973). Discrimination of mirror images by young children. *Journal of Comparative Physiological Psychology, 82*, 415–425.

Bryant, P. E. (1969). Perception and memory of the orientation of visually presented lines by children. *Nature, 224*, 1331–1332.

Burton, L. (2004). *Mathematicians as enquirers: learning about learning mathematics*. Dordrecht: NL: Springer.

Busch, J. C., Watson, J. A., Brinkley, V., Howard, J. R., & Nelson, C. (1993). Preschool embedded figures test performance of young children: age and gender differences. *Perceptual and Motor Skills, 77*(2), 491–496.

Cakmak, S., Isiksal, M., & Koc, Y. (2014). Investigating effect of origami-based instruction on elementary students' spatial skills and perceptions. *The Journal of Educational Research, 107*(1), 59–68.

Campbell, S. R. (2010). Embodied minds and dancing brains: new opportunities for research in mathematics education. In B. Sriaman & L. English (Eds.), *Theories of mathematics education* (pp. 309–331). Berlin: Springer Berlin.

Campbell, S. R. (2011). Educational neuroscience: motivations, methodology, and implications. In K. E. Patten & S. R. Campbell (Eds.), *Educational neuroscience: initiatives and emerging issues* (pp. 7–16). Hoboken, NJ: Wiley-Blackwell.

Carter, N. (2009). *Visual group theory*. New York: Mathematical Association of America.

Case, R., Stephenson, K. M., Bleiker, C., & Okamoto, Y. (1996). Central spatial structures and their development. In R. Case & Y. Okamoto (Eds.), The role of central conceptual structures in the development of children's thought. *Monographs of the Society for Research in Child Development, 61*(1–2), 83–102. Serial 246.

Casey, B., Andrews, N., Schindler, H., Kersh, J. E., Samper, A., & Copley, J. (2008). The development of spatial skills through interventions involving block building activities. *Cognition and Instruction, 26*(3), 269–309.

Casey, B., Erkut, S., Ceder, I., & Young, J. (2008). Use of a storytelling context to improve girls' and boys' geometry skills in Kindergarten. *Journal of Applied Developmental Psychology, 29*, 29–48.

Caswell, B., Moss, J., Hawes, Z., & Naqvi, S. (2013). Math for young children (M4YC) project: a no-ceiling approach to math learning in an urban school.

Centre for Urban Schooling, 1(1), 1–4.

Châtelet, G. (2000/1993). Les enjeux du mobile. Paris: Seuil. (Engl. transl., by R. Shore & M. Zagha: Figuring space: Philosophy, Mathematics and Physics). Dordrecht: Kluwer Academy Press.

Cheng, K., Huttenlocher, J., & Newcombe, N. S. (2013). 25 years of research on the use of geometry in spatial reorientation: A current theoretical perspective. Psychonomic Bulletin & Review, 1–22.

Cheng, Y.-L., & Mix, K. S. (2014). Spatial training improves children's mathematics ability. Journal of Cognition and Development, 15(1), 2–11. doi: 10.1080/15248372.2012.725186

Chu, M., & Kita, S. (2011). The nature of gestures' beneficial role in spatial problem solving. Journal of Experimental Psychology: General, 140(1), 102–116.

Claessens, A., Duncan, G., & Engel, M. (2009). Kindergarten skills and fifth-grade achievement: evidence from the ECLS-K. Economics of Education Review, 28, 415–427.

Claessens, A., & Engel, M. (2011). How important is where you start? Early mathematical knowledge and later school success. Paper presented at the 2011 Annual Meeting of the American Educational Research Association (AERA), New Orleans, LA. April 2011.

Clearfield, M. W. (2004). The role of crawling and walking experience in infant spatial memory. Journal of Experimental Child Psychology, 89(3), 214–241.

Clements, D. H. (2004). Geometric and spatial thinking in early childhood education. In D. H. Clements, J. Sarama, & A.-M. Di Biase (Eds.), Engaging young children in mathematics: Standards for early childhood mathematics education (pp. 267–298). Mahwah, NJ: Lawrence Earlbaum Associates, Inc.

Clements, D. H., & Sarama, J. (2004). Mathematics everywhere, every time. Teaching Children Mathematics, 10(8), 421–426.

Clements, D. H., & Sarama, J. (2011). Early childhood teacher education: the case of geometry. Journal of Mathematics Teacher Education, 14(2), 133–148.

Clements, D. H., Swaminathan, S., Hannibal, M. A. Z., & Sarama, J. (1999). Young children's concepts of shape. Journal for Research in Mathematics Education, 30(2), 192–212.

Clements, D. H., Wilson, D. C., & Sarama, J. (2004). Young children's composition of geometric figures: A learning trajectory. Mathematical Thinking and Learning, 6(2), 163–184.

Clements, M. A. (1982). Visual imagery and school mathematics. For the Learning of Mathematics, 2, 2–9, & 3, 33–39.

Clements, M. A. (2008). Spatial abilities, mathematics, culture, and the Papua New Guinea experience. In P. Clarkson & N. Presmeg (Eds.), Critical issues in mathematics education (pp. 97–106). New York, NY: Springer. Retrieved from http://link.springer.com.ezproxy.lib.ucalgary.ca/chapter/10.1007/978-0-387-09673-5_7

Coates, S. W. (1972). Preschool embedded figures test. Palo Alto, CA: Consulting Psychologists Press.

Cohn, N. (2014). Framing "I can't draw": The influence of cultural frames on the development of drawing. Culture and Psychology, 20(1), 102–117.

Cook, S. W., Mitchell, Z. & Goldin-Meadow, S. (2008). Gesturing makes learning last. Cognition, 106(2), 1047–1058.

Coxeter, H. S. M. et al. (1967). Geometry Curriculum K–13. Toronto: OISE/UofT 1967. http://wiki.math.yorku.ca/index.php/CMEF_Geometry_Curriculum

Cutting, J. E. (2002). Representing motion in a static image: constraints and parallels in art, science, and popular culture. *Perception, 31,* 1165–1193.

D'Ambrosio, U. (1985). Ethnomathematics and its place in the history and pedagogy of mathematics. *For the Learning of Mathematics, 5*(1), 44–48.

Davis, B. (1996). *Teaching mathematics: toward a sound alternative.* New York: Garland.

Davis, B. (2011). Mathematics teachers' subtle, complex disciplinary knowledge. *Science, 332,* 1506–1507.

Davis, B., Sumara, D., & Luce-Kapler, R. (2008). *Engaging minds: changing teaching in complex times (2nd ed.).* New York: Routledge.

de Freitas, E., & Sinclair, N. (2012). Diagram, gesture, agency: theorizing embodiment in the mathematics classroom. *Educational Studies in Mathematics, 80*(1–2), 133–152.

de Freitas, E., & Sinclair, N. (2013). New materialist ontologies in mathematics education: the body in/of mathematics. *Educational Studies in Mathematics, 83*(3), 453–470.

de Hevia, M. D., Vallar, G., & Girelli, L. (2008). Visualizing numbers in the mind's eye: the role of visuo-spatial process in numerical abilities. *Neuroscience and Biobehavioral Reviews, 32*(8), 1361–1372.

Dehaene, S. (2011). *The number sense: how the mind creates mathematics,* revised edition. Oxford, UK: Oxford University Press.

Dehaene, S., Bossini, S., & Giraux, P. (1993). The mental representation of parity and number magnitude. *Journal of Experimental Psychology: General, 122*(3), 371–396.

DeLoache, J. S. (1987). Rapid change in the symbolic functioning of very young children. *Science, 238*(4833), 1556–1557.

Diamond, J. (2012). *The world until yesterday: what can we learn from traditional societies?* New York: Penguin Books.

Diefes-Dux, H. A., Whittenberg, L., & McKee, R. (2013). Mathematical modeling at the intersection of elementary mathematics, art, and engineering education. In L. D. English & J. T. Mulligan (Eds.), *Reconceptualizing early mathematics learning* (pp. 309–325). Dordrecht, NL: Springer.

Dieudonné, J. (1981). The universal domination of geometry. *The Two-Year College Mathematics Journal, 12*(4), 227–231.

Donaldson, M. (1986). *Children's explanations: a psycholinguistic study.* Cambridge, UK: Cambridge University Press

Doyle, R. A., Voyer, D., & Cherney, I. D. (2012). The relation between childhood spatial activities and spatial abilities in adulthood. *Journal of Applied Developmental Psychology, 33*(2), 112–120.

Duncan, G. J., Dowsett, C. J., Claessens, A., Magnuson, K., Huston, A.C., Klebanov, P., & Japel, C. (2007). School readiness and later achievement. *Developmental Psychology, 43,* 1428–1446.

Dreyfus, T. (1994). The role of cognitive tools in mathematics education. In R. Biehler, R. Scholz, R. Strasser, & B. Winkelmann (Eds.), *Didactics of mathematics as a scientific discipline* (pp. 201–212). Dordrecht: Kluwer.

Edwards, L., Nathan, M., Nemirovsky, R., & Soto-Johnson, H. (2013). *Embodied cognition: what it means to know and do mathematics.* Symposium conducted at the Annual Meeting of the National Council of Teachers of Mathematics (NCTM) Research Pre-session, Denver, CO.

Eglash, R. (1999). *African fractals: modern computing and indigenous design.* New Brunswick, NJ: Rutgers University Press.

Ekstrom, R. B., French, J. W., Harman, H. H., & Dermen, D. (1976). *Kit of factor-referenced cognitive tests.* Princeton, NJ: Educational Testing Service.

Ehrlich, S., Levine, S., & Goldin-Meadow, S. (2006). The importance of gesture in children's spatial reasoning. *Developmental Psychology, 42*(6), 1259–1268.

Empson, S. B., & Turner, E. (2006). The emergence of multiplicative thinking in children's solutions to paper folding tasks. *The Journal of Mathematical Behavior, 25*(1), 46–56.

Farmer, G., Verdine, B., Lucca, K., Davies, T., Dempsey, R., Newcombe, N., Hirsh-Pasek, K., & Golinkoff, R. (2013). *Putting the pieces together: spatial skills at age 3 predict to spatial and math performance at age 5.* Paper presented at the Society for Research in Child Development, Seattle, WA.

Fauconnier, G., & Turner, M. (2002). *The way we think: conceptual blending and the mind's hidden complexities.* New York: Basic Books.

Feng, J., Spence, I., & Pratt, J. (2007). Playing an action video game reduces gender differences in spatial cognition. *Psychological Science, 18*(10), 850–855. doi: 10.1111/j.1467-9280.2007.01990.x

Fias, W., Menon, V., & Szucs, D. (2013). Multiple components of developmental dyscalculia. *Trends in Neuroscience and Education, 2,* 43–47.

Fischer, M. H., & Shaki, S. (2014). Spatial associations in numerical cognition: from single digits to arithmetic. *Quarterly Journal of Experimental Psychology, 67,* 1461–1483. doi: 10.1080/17470218.2014.927515

Flavell, J. H. (1999). Cognitive development: children's knowledge about the mind. *Annual Review of Psychology, 50*(1), 21–45.

Fomenko, A. T. (2011). *Visual geometry and topology.* Berlin: Springer.

Fonagy, P., & Target, M. (2007). The rooting of the mind in the body: new links between attachment theory and psychoanalytic thought. *Journal of American Psychoanalytic Association, 55*(2), 411–456. doi: 10.1111/j.1469-7610.2007.01727.x

Frick, A., Daum, M. M., Walser, S., & Mast, F. W. (2009). Motor processes in children's mental rotation. *Journal of Cognition and Development, 10,* 18–40.

Frick, A., Möhring, W., & Newcombe, N. S. (2014). Picturing perspectives: development of perspective-taking abilities in 4- to 8-year-olds. *Frontiers in Psychology, 5*:386. doi: 10.3389/fpsyg.2014.00386

Frick, A., Möhring, W., & Newcombe, N. S. (in press). Development of mental transformation abilities. *Trends in Cognitive Sciences.* doi: http://dx.doi.org/10/1016/j.tics.2014.05.011

Frick, A., & Newcombe, N. S. (2012). Getting the big picture: development of spatial scaling abilities. *Cognitive Development, 27*(3), 270–282.

Frick, A., & Wang, S. H. (2014). Mental spatial transformations in 14- and 16-month-old infants: effects of action and observational experience. *Child Development, 85*(1), 278-293. doi: 10.1111/cdev.12116

Galton, F. (1880). Visualised numerals. *Nature, 21,* 252–256.

Galton, F. (1881). Visualised numerals. *Journal of the Anthropological Institute of Great Britain and Ireland,* 10: 85–102.

Gattegno, C. (1963). *For the teaching of mathematics.* Reading, UK: Educational Explores.

Geary, D. C. (2004). Mathematics and learning disabilities. *Journal of Learning Disabilities, 37*(4), 4–15.

Geary, D. C., Hoard, M. K., Byrd-Craven, J., Nugent, L., & Numtee, C. (2007). Cognitive mechanisms underlying achievement deficits in children with mathematical learning disability. *Child Development, 78*(4), 1343–1359.

Gerofsky, S. (2014). Making sense of multiple meanings of "embodied mathematics learning." In S. Oesterle, P. Liljedahl, C. Nicol, & D. Allan (Eds.), *Proceedings of the Joint Meeting of PME 38 and PME-NA 36*, vol. 3, pp. 145–152. Vancouver, Canada: PME.

Girelli, L., Semenza, C., & Delazer, M. (2004). Inductive reasoning and implicit memory: evidence from intact and impaired memory systems. *Neuropsychologia, 42*(7), 926–938.

Gray, H. (1858). *Anatomy of the human body*. New York: Lea and Febiger.

Gregory, K. M., Kim, A. S., & Whiren, A. (2003). The effect of verbal scaffolding on the complexity of preschool children's block constructions. In D. E. Lytle (Ed.), *Play and educational theory and practice: play and culture studies* (vol. 5, pp. 117–133). Westport, CT: Praeger.

Grissmer, D. W., Mashburn, A. J., Cottone, E., Chen, W. B., Brock, L. L., Murrah, W. M., & Cameron, C. E. (2013). *Play-based after-school curriculum improves measures of executive function, visuospatial and math skills and classroom behavior for high risk K–1 children*. Paper presented at the Society for Research in Child Development, Seattle, WA.

Grumet, M. (1988). *Bitter milk: women and teaching*. Amherst, MA: University of Massachusetts Press.

Gunderson, E. A., Ramirez, G., Beilock, S. L., & Levine, S. C. (2012). The relation between spatial skills and early number knowledge: the role of the linear number line. *Developmental Psychology, 48*(5), 1229–1241. doi: 10.1037/a0027433

Hallowell, D. A., Okamoto, Y., Romo, L. F., & La Joy, J. R. (in press). First-grader's spatial-mathematical reasoning about plane and solid shapes and their representations. *ZDM: The International Journal on Mathematics Education*.

Hanline, M. F., Milton, S., & Phelps, P. (2001). Young children's block construction activities: findings from 3 years of observation. *Journal of Early Intervention, 24*, 224–237.

Happé, F. (2013). Embedded figures test (EFT). In F. R. Volkmar (Ed.), *Encyclopedia of autism spectrum disorders* [electronic resource] (pp. 1077–1078). New York: Springer.

Harris, J., Hirsh-Pasek, K., & Newcombe, N. S. (2013). A new twist on studying the development of dynamic spatial transformations: Mental paper folding in young children. *Mind, Brain and Education, 7*(1), 49–55.

Hawes, Z., Chang, D., Naqvi, S., Olver, A., & Moss, J. (2013). Uncovering the processes of young children's 3D mental rotation abilities: implications for lesson design. In M. Martinez & A. Castro Superfine (Eds.), *Proceedings of the 35th annual meeting of the North American Chapter of the International Group for the Psychology of Mathematics Education*. Chicago, IL: University of Illinois at Chicago.

Hawes, Z., Lefevre, J., Xu, C., & Bruce, C. (in press). Mental rotation with tangible three-dimensional objects: a new measure sensitive to developmental differences in 4- to 8-year-old children. *Mind, Brain and Education, 8*(4).

Hawes, Z., Moss, J., Caswell, B., Naqvi, S., & MacKinnon (in preparation). Enhancing children's spatial and numerical skills through a 'dynamic-spatial' approach to early geometry instruction: effects of a seven-month intervention.

Hawkins, D. (2000). *The roots of literacy*. Boulder, CO: University Press of Colorado.

Healy, L., & Fernandes, S. H. A. A. (2011). The role of gestures in the mathematical practices of those who do not see with their eyes. *Educational Studies in*

Mathematics, 77, 157–174.

Hegarty, M., Crookes, R. D., Dara-Abrams, D., & Shipley, T. F. (2010). Do all science disciplines rely on spatial abilities? Preliminary evidence from self-report questionnaires. *Spatial Cognition, VII.* 85–94.

Henderson, D. W., & Taimina, D. (2005). *Experiencing geometry: Euclidean and non-Euclidean with history,* New York: Prentice Hall.

Henrich, J., Heine, S. J., & Norenzayan, A. (2010). The weirdest people in the world? *Behavioral and Brain Sciences, 33*(2–3), 61–83. doi: 10.1017/S0140525X0999152X

Hershkowitz, R., Parzysz, B., & Van Dormolen, J. (1996). Space and shape. In A. Bishop, K. Clements, C. Keitel, J. Kilpatrick, & C. Laborde (Eds.), *International Handbook of Mathematics Education* (pp. 61–204). Dordrecht: Kluwer.

Hiley, D. R., & Bohman, J. F. (1992). *The interpretive turn: philosophy, culture, science.* Ithaca, NY: Cornell University Press.

Hodgen, J., Pepper, D., Sturman, L., & Ruddick, G. (2010). *Is the UK an outlier? An international comparison of upper secondary mathematics education.* London: Nuffield Foundation.

Hoffman, D. (2000a). *Visual intelligence: how we create what we see (new edition).* New York: W.W. Norton.

Hoffman, D. (2000b). *When the world stopped moving.* Animated gifs from http://www.cogsci.uci.edu/%7Eddhoff/vi6.html

Holmes, J., Gathercole, S. E., & Dunning, D. L. (2009). Adaptive training leads to sustained enhancement of poor working memory in children. *Developmental Science, 12*(4), 9–15.

Howson, A. G. (Ed.). (1973). *Developments in mathematical education: Proceedings of the Second International Congress on Mathematical Education.* Cambridge, UK: Cambridge University Press.

Hubbard, E. M., Piazza, M., Pinel, P., & Dehaene, S. (2009). Numerical and spatial intuitions: a role for posterior parietal cortex? In L. Tommasi, L. Nadel, & M. A. Peterson (Eds.), *Cognitive biology: evolutionary and developmental perspectives on mind, brain and behavior* (pp. 221–246). Cambridge, MA: MIT Press.

Huttenlocher, J., Newcombe, N. S., & Vasilyeva, M. (1999). Spatial scaling in young children. *Psychological Science, 10,* 393–398.

Jardine, D. W. (1994). On the ecologies of mathematical language and the rhythms of the earth. In P. Ernest (Ed.), *Mathematics, education, and philosophy: an international perspective* (pp. 109–23), London, UK: Falmer.

Jaušovec, N., & Jaušovec, K. (2012). Sex differences in mental rotation and cortical activation patterns: Can training change them? *Intelligence, 40*(2), 151–162.

Jmol: an open-source Java viewer for chemical structures in 3D. http://www.jmol.org/

Karp, S. A., & Konstadt, N. (1971). *Children's Embedded Figures Test.* Palo Alto, CA: Consulting Psychologists Press.

Kastens, K. A., & Ishikawa, T. (2006). Spatial thinking in the geosciences and cognitive sciences: a cross-disciplinary look at the intersection of the two fields. In C. A. Manduca & D. W. Mogk (Eds.), *Earth and mind: How geologists think and learn about the Earth* (pp. 53–76). Boulder, CO: Geological Society of America.

Kell, H. J., Lubinski, D., Benbow, C. P., & Steiger, J. H. (2013). Creativity and technical innovation: spatial ability's unique role. *Psychological Science, 24*(9), 1831–1836. doi: 10.1177/0956797613478615.

Kirkwood, M. W., Weiler, M. D, Bernsetin, J. H., Forbes, P. W., & Waber, D. P. (2001). Sources of poor performance on the Rey-Osterrieth Complex Figure Test among children with learning difficulties: a dynamic assessment approach.

The Clinical Neuropsychologist, 15(3), 345–356.

Kotsopoulos, D., Cordy, M., & Langemeyer, M. (under review). Children's understanding of large-scale mapping tasks: an analysis of talk, drawings, and gesture. *ZDM: The International Journal on Mathematics Education.*

Kotsopoulos, D., Cordy, M., Langemeyer, M., & Khattak, L. (2014). How do children draw, describe, and gesture about motion? In *the proceedings of the 11th International Conference of the Learning Sciences* (pp. 1–2). Boulder, Colorado.

Kozhevnikov, M., Hegarty, M., & Mayer, R. E. (2002). Revising the visualizer-verbalizer dimension: evidence for two types of visualizers. *Cognition and Instruction, 20*(1), 47–77.

Kozhevnikov, M., Kosslyn, S., & Shephard, J. (2005). Spatial versus object visualizers: a new characterization of visual cognitive style. *Memory & Cognition, 33*(4), 710–726.

Krüger, M., Kaiser, M., Mahler, K., Bartels, W., & Krist, H. (2014). Analogue mental transformations in 3-year-olds: introducing a new mental rotation paradigm suitable for young children. *Infant and Child Development, 23*(2), 123–138. doi: 10.1002/icd.1815

Kyttälä, M., Aunio, P., Lehto, J. E., Van Luit, J., & Hautamaki, J. (2003). Visuospatial working memory and early numeracy. *Educational and Child Psychology, 20*(3), 65–76.

Kyttälä, M., & Lehto, J. (2008). Some factors underlying mathematical performance: the role of visuospatial working memory and non-verbal intelligence. *European Journal of Psychology of Education, XXII*(1), 77–94.

Lakoff, G., & Johnson, M. (1999). *Philosophy in the flesh: the embodied mind and its challenge to western thought.* New York: Basic Books.

Lakoff, G., & Núñez, R. (2000). *Where mathematics comes from: how the embodied mind brings mathematics into being.* New York: Basic Books.

Lee, S. A., Sovrano, V. A., & Spelke, E. S. (2012). Navigation as a source of geometric knowledge: young children's use of length, angle, distance, and direction in a reorientation task. *Cognition, 123*(1), 144–161.

Leikin, R., Berman, A., & Zaslavsky, O. (2000). Applications of symmetry to problem solving. *International Journal of Mathematical Education in Science and Technology, 31*, 799–809.

Levine, S. C., Ratcliff, K. R., Huttenlocher, J., & Cannon, J. (2012). Early puzzle play: a predictor of preschooler's spatial transformation skill. *Developmental Psychology, 48*, 530–542.

Linn, M. C., & Petersen, A. C. (1985). Emergence and characterization of sex differences in spatial ability: a meta-analysis. *Child Development, 56*, 1479–1498.

Lowrie, T. (2002). The influence of visual and spatial reasoning in interpreting simulated 3D worlds. *International Journal of Computers for Mathematical Learning, 7*(3), 301–318. doi: 10.1023/A:1022116221735

Malafouris, L. (2013). *How things shape the mind: a theory of material engagement.* Cambridge, MA: The MIT Press.

Malisza, K. L., Clancy, C., Shiloff, D., Foreman, D., Holden, J., Jones, C., Paulson, K., Summers, R., Yu, C. T., & Chudley, A. E. (2011). Functional evaluation of hidden figures object analysis in children with autistic disorder. *Journal of Autism Developmental Disorders, 41*, 13–22.

Mamolo, A., Sinclair, M., & Whiteley, W. (2011). Filling the pyramid: an activity in proportional reasoning. *Mathematics Teaching in the Middle School, 16*, 545–551.

Mamolo, A., & Whiteley, W. (2012). The Popcorn Box activity and reasoning about optimization. *Mathematics Teacher*, *105*(6), 420–426.

Manley, D. B., & Taylor, C. S. (1996). *Descartes' meditations*. Retrieved from http://www.wright.edu/cola/descartes/synopsis.html

Martin, L. C., & Towers, J. (2011). Improvisational understanding in the mathematics classroom. In R. Keith Sawyer (Ed.), *Structure and improvisation in creative teaching* (pp. 252–278). New York: Cambridge University Press.

Martin, L., Towers, J., & Pirie, S. (2006). Collective mathematical understanding as improvisation. *Mathematical Thinking and Learning*, *8*(2), 149–183.

Maturana, H. R., & Varela, F. J. (1972). *Autopoiesis and cognition: the realization of the living*. Dordrecht, NL: Reidel.

Maturana, H. R., & Varela, F. J. (1991). *The tree of knowledge: the biological roots of human understanding* (Rev. ed.). Boston, MA: Random House. (Original work published 1987.)

Mazzocco, M. M., & Räsänen, P. (2013). Contributions of longitudinal studies to evolving definitions and knowledge of developmental dyscalculia. *Trends in Neuroscience and Education*, *2*(2), 65–73. doi: 10.1016/j.tine.2013.05.001

McAuliffe, C. (2003). *Visualizing topography: effects of presentation strategy, gender, and spatial ability* (Ph.D. Doctoral dissertation). Arizona State University, Dissertation Abstracts International, Volume: 64-10, Section: A, page: 3653. Retrieved from http://adsabs.harvard.edu/abs/2003PhDT.......162M (AAI3109579)

Menghini, M., Furinghetti, F., Giacardi, L., & Arzarello, F. (Eds.). (2008). *The first century of the International Commission on Mathematical Instruction (1908–2008): Reflecting and shaping the world of mathematics education*. Rome: Instituto della Enciclopedia Italiana.

Merleau-Ponty, M. (1945). *Phénoménologie de la perception*. Paris: Gallimard.

Merleau-Ponty, M. (1962). *Phenomenology of perception*. (C. Smith, Trans.) New York: Routledge Classics. (Original work published 1945).

Merleau-Ponty, M. (1964). *Le visible et l'invisible* [The visible and invisible]. Paris: Gallimard.

Mix, K. S., & Cheng, Y. L. (2012). The relation between space and math: developmental and educational implications. In J. B. Benson (Ed.), *Advances in child development and behavior* (vol. 42, pp. 197–243). San Diego, CA: Academic Press.

Möhring, W., Newcombe, N. S., & Frick, A. (2014). Zooming in on spatial scaling: preschool children and adults use mental transformations to scale spaces. *Developmental Psychology*, *50*(5), 1614–1619.

Moll, H., Meltzoff, A. N., Merzsch, K., & Tomasello, M. (2013). Taking versus confronting visual perspectives in preschool children. *Developmental Psychology*, *49*(4), 646.

Moss, J., Hawes, Z., Chang, D., & Tepylo, D. (2013). *From mental rotation task design to lesson planning: the case of the polyomino challenge*. Paper presented at Fields MathEd Forum Meeting, Toronto, ON: The Fields Institute for Research in Mathematical Sciences, University of Toronto.

Moss, J., Hawes, Z., Naqvi, S., & Caswell, B. (in press). Adapting Japanese Lesson Study to enhance the teaching and learning of geometry and spatial reasoning in early years classrooms: a case study. *ZDM: The International Journal on Mathematics Education*, *47*(3).

Mulligan, J. T., & Mitchelmore, M. C. (2009). Awareness of pattern and structure in early mathematical development. *Mathematics Education Research Journal*,

21(2), 33–49.

Naqvi, S., Hawes, Z., Chang, D., & Moss, J. (2013). Exploring pentominoes in 7 diverse pre-K/K classrooms. In M. Martinez & A. Castro Superfine (Eds.), *Proceedings of the 35ᵗʰ annual meeting of the North American Chapter of the International Group for the Psychology of Mathematics Education*. Chicago, IL: University of Illinois at Chicago.

NASA. (2008). *Amino Acid Chirality*. Retrieved from http://commons.wikimedia.org/wiki/File:Chirality_with_hands.jpg

Nath, S., & Szücs, D. (2014). Construction play and cognitive skills associated with the development of mathematical abilities. *Learning and Instruction, 32*, 73–80.

Nathan, M. J., Eilam, B., & Kim, S. (2007). To disagree, we must also agree: how intersubjectivity structures and perpetuates discourse in a mathematics classroom. *The Journal of the Learning Sciences, 16*(4), 523–563.

National Council of Teachers of Mathematics. (2010). *Focus in Grade 1: Teaching with Curriculum Focal Points*. Reston, VA: National Council of Teachers of Mathematics.

National Research Council. (2006). *Learning to think spatially: GIS as a support system in the K–12 curriculum*. Washington, DC, The National Academies Press.

Needham, T. (2004). *Visual complex analysis*. New York: Clarendon Press.

Nemirovsky, R., Borba, M., Dimattia, C., Arzarello, F., Robutti, O., Schnepp, M., Chazan, D., Ramussen, C., Olszewski, J., Dost, K., Johnson, J. L., Borba, M. C., & Scheffer, N. F. (2004). PME Special Issue: Bodily activity and imagination in mathematics learning. *Educational Studies in Mathematics, 57*(3), 303–321.

Nemirovsky, R., & Ferrara, F. (2009). Mathematical imagination and embodied cognition. *Educational Studies in Mathematics, 70*(2), 159–174.

Nemirovsky, R., Kelton, M. L., & Rhodehamel, B. (2013). Playing mathematical instruments: emerging perceptuomotor integration with an interactive mathematics exhibit. *Journal for Research in Mathematics Education, 44*(2), 372–415.

Newcombe, N. S. (2010). Picture this: increasing math and science learning by improving spatial thinking. *American Educator, 34*(2), 29–43.

Newcombe, N. S. (2014). *Thinking about quantity: the intertwined development of spatial and numerical cognition*. Paper presented at the Workshop on Making Models, Toronto, ON.

Newcombe, N., Huttenlocher, J., Drummey, A. B., & Wiley, J. G. (1998). The development of spatial location coding: place learning and dead reckoning in the second and third years. *Cognitive Development, 13*(2), 185–200.

Newcombe, N. S., & Shipley, T. F. (in press). Thinking about spatial thinking: new typology, new assessments. In J. S. Gero (Ed.), *Studying visual and spatial reasoning for design creativity*. New York: Springer.

Newcombe, N. S., & Stieff, M. (2012). Six myths about spatial thinking. *International Journal of Science Education, 34*(6), 955–971.

Nosworthy, N., Bugden, S., Archibald, L., Evans, B., & Ansari, D. (2013). A two-minute paper-and-pencil test of symbolic and nonsymbolic numerical magnitude processing explains variability in primary school children's arithmetic competence. *PLoS ONE 8*(7): e67918. doi: 10.1371/journal.pone.0067918

Núñez, R. (2003). Do real numbers really move? Language, thought, and gesture: the embodied cognitive foundations of mathematics. In R. Hersh (Ed.), *18 unconventional essays on the nature of mathematics* (pp. 160–181). New York: Springer.

Núñez, R. E., Edwards, L. D., & Matos, J. F. (1999). Embodied cognition as grounding for situatedness and context in mathematics education. *Educational Studies in Mathematics, 39*(1–3), 45–65.

Okagaki, L., & Frensch, P. A. (1994). Effects of video game playing on measures of spatial performance: gender effects in late adolescence. *Journal of Applied Developmental Psychology, 15*(1), 33–58.

Ong, W. (1982). *Orality and literacy: the technologizing of the word.* New York: Routledge.

Ontario Ministry of Education and Training/OMET. (2005). *The Ontario curriculum Grades 1–8 mathematics, Revised.* Toronto: Queen's Printer for Ontario.

Ormand, C. J., Manduca, C., Shipley, T. F., Tikoff, B., Harwood, C. L., Atit, K., & Boone, A. P. (2014). Evaluating geoscience students' spatial thinking skills in a multi-institutional classroom study. *Journal of Geoscience Education, 62*(1), 146–154. doi: 10.5408/13-027.1

Ormand, C. J., Shipley, T. F., Tikoff, B., Manduca, C. A., Dutrow, B., Goodwin, L., & Resnick, I. (2013, June). *Improving spatial reasoning skills in the undergraduate geoscience classroom through interventions based on cognitive science research.* Presented at the AAPG Hedberg Conference on 3D Structural Geologic Interpretation, Reno, Nevada. Retrieved from http://spatiallearning.org/media/silc_pdfs/resources/testsandinstruments/tandi-images/crystal_slicing_test.jpg

Pascual-Leone, J., & Morra, S. (1991). Horizontality of water level: a neo-piagetian developmental review. In H. W. Reese (Ed.), *Advances in Child Development and Behavior* (Vol. 23, pp. 231–276). New York: Academic Press.

Piaget, J., & B. Inhelder (1967/1948). *The child's conception of space.* (F. J. Langdon & J. L. Lunzer, Trans.). New York: Norton.

Piaget, J., Inhelder, B., & Szeminska, A. (1960). *The child's conception of geometry.* London: Routledge and Kegan Paul.

Pickering, A. (1995). *The mangle of practice: time, agency and science.* Chicago: University of Chicago Press.

Pirie, S., & Kieren, T. (1994). Growth in mathematical understanding: how can we characterise it and how can we represent it? *Educational Studies in Mathematics, 26*(2–3), 165–190.

Pirie, S. E. B., & Thom, J. S. (2001). Thinking through ecological metaphors in mathematics education. In S. Gunn & A. Begg (Eds.), *Mind, body, and society: emerging understandings of knowing and learning* (pp. 45–52). Melbourne, AU: University of Melbourne.

Poirier, L. (2007). Teaching mathematics and the Inuit community. *Canadian Journal of Science, Mathematics, & Technology Education, 7*(1), 53–67.

Presmeg, N. C. (1986). Visualization in high school mathematics. *For the Learning of Mathematics, 6*(3), 42–46.

Presmeg, N. (2008). Spatial abilities research as a foundation for visualization in teaching and learning mathematics. In P. Clarkson & N. Presmeg (Eds.), *Critical Issues in Mathematics Education* (pp. 83–95). New York: Springer.

Pruden, S. M., Levine, S. C., & Huttenlocher, J. (2011). Children's spatial thinking: does talk about the spatial world matter? *Developmental Science, 14*(6), 1417–1430. doi: 10.1111/j.1467-7687.2011.01088.x

Pulvermüller, F. (2011). Meaning and the brain: the neurosemantics of referential, interactive, and combinatorial knowledge. *Journal of Neurolinguistics, 25*(5), 423–459.

Quaiser-Pohl, C., Lehmann, W., & Eid, M. (2004). The relationship between spatial abilities and representations of large-scale space in children – a structural equation modeling analysis. *Personality and Individual Differences, 36*(1), 95–107.

Quinn, P. C., Siqueland, E. R., & Bomba, P. C. (1985). Delayed recognition memory for orientation by human infants. *Journal of Experimental Child Psychology, 40*(2), 293–303.

Radford, L. (2002). The seen, the spoken and the written: a semiotic approach to the problem of objectification of mathematical knowledge. *For the Learning of Mathematics, 22*(2), 1423.

Radford, L. (2014). Towards an embodied, cultural, and material conception of mathematics cognition. *ZDM: The International Journal on Mathematics Education, 46,* 349–361.

Radford, L., Demers, S., Guzmán, J., & Cerulli, M. (2003). Calculators, graphs, gestures and the production of meaning. In *Proceedings of the 27th Conference of the International Group for the Psychology of Mathematics Education* (pp. 55–62). Honolulu, HI: University of Hawaii.

Radford, L., Edwards, L., & Arzarello, F. (2009). Introduction: beyond words. *Educational Studies in Mathematics, 70*(2), 91–95.

Ramey, J., & Uttal, D. H. (2014). Distributed spatial sensemaking in middle school engineering learning, SILC Showcase, November 2014, http://bit.ly/1sNA3OJ

Reifel, S., & Greenfield, P. M. (1983). Part-whole relations: some structural features of children's representational block play. *Child Care Quarterly, 12*(2), 144–151.

Rieser, J. J., Garing, A. E., & Young, M. E. (1994). Imagery, action, and young children's spatial orientation: It's not being there that counts, it's what one has in mind. *Child Development, 65,* 1262–1278.

Robert, M., & Héroux, G. (2003). Visuo-spatial play experience: forerunner of visuo-spatial achievement in preadolescent and adolescent boys and girls. *Infant and Child Development, 13*(1), 49–78.

Roth, W.-M. (2001). Gestures: their role in teaching and learning. *Review of Educational Research, 71,* 365–392.

Ruddick, G., & Sainsbury, M. (2008). *Comparison of the core primary curriculum in England and those of other high performing countries.* London: National Foundation for Educational Research.

Russell, B. (1903/2010). *Principles of mathematics.* London: Routledge.

Sacks, O. (1993, May 10). A Neurologist's Notebook: To see and not see. *The New Yorker,* pp. 59–73.

Sagiv, N., Simner, J., Collins, J., Butterworth, B., & Ward, J. (2006). What is the relationship between synaesthesia and visuo-spatial number forms? *Cognition, 101,* 114–128.

St. Clair-Thompson, H. L., Stevens, R., Hunt, A., & Bolder, E. (2010). Improving children's working memory and classroom performance. *Educational Psychology, 30,* 203–220.

Sarama, J., & Clements, D. H. (2009). *Early childhood mathematics education research: Learning trajectories for young children.* New York: Routledge.

Sawyer, W. W. (2003). *Vision in elementary mathematics.* London: Courier Dover Books in Mathematics.

Schubring, G. (Ed.). (n.d.). *International Journal for the History of Mathematics Education.* New York: Teachers College Press. Available from http://www.

comap.com/historyjournal

Seo, K.-H., & Ginsburg, H. P. (2004). What is developmentally appropriate in early childhood mathematics education? Lessons from new research. In D. H. Clements, J. Sarama, & A.-M. DiBiase (Eds.), *Engaging young children in mathematics: standards for early childhood mathematics education* (pp. 91–104). Hillsdale, NJ: Erlbaum.

Seron, X., Pesenti, M., Noël, M.-P., Deloche, G., & Cornet, J.-A. (1992). Images of numbers, or "when 98 is upper left and 6 sky blue." *Cognition, 44*(1–2), 159–196.

Seung, H. S., & Lee, D. D. (2000). The manifold ways of perception. *Science, 290*(5500), 2268–2269.

Shaki, S., Fischer, M. H., & Petrusic, W. M. (2009). Reading habits for both words and numbers contribute to the SNARC effect. *Psychonomic Bulletin and Review, 16*(2), 328–331.

Sheets-Johnstone, M. (2012). Movement and mirror neurons: a challenging and choice conversation. *Phenomenology and the Cognitive Sciences, 11*(3), 385–401.

Shepard, R. N., & Metzler, J. (1971). Mental rotation of three-dimensional objects. *Science, 171*, 701–703.

Sherin, B. I. (2001). How students invent representations of motion: a genetic account. *Journal of Mathematical Behavior, 19*, 399–441.

Siegler, R. S., & Booth, J. L. (2004). Development of numerical estimation in young children. *Child Development, 75*, 428–444.

Siegler, R. S., & Ramani, G. B. (2008). Playing linear numerical board games promotes low-income children's numerical development. *Developmental Science, 11*, 655–661.

Siegler, R. S., & Ramani, G. B. (2009). Playing linear number board games – but not circular ones – improves low-income preschoolers' numerical understanding. *Journal of Educational Psychology, 101*, 545–560.

SIGGRAPH. (2002). *White Paper on Visual Learning in Science and Engineering, report from the Visual Learning Campfire*, Snowbird, UT: ACMSIGGRAPH Education Committee, http://education.siggraph.org/conferences/other/visual-learning

Sinclair, N. (2008). *The history of geometry curriculum in the United States*. Charlotte, NC: Information Age Publishing.

Sinclair, N. (2014). *Motion in mathematics*. Keynote address at the British Columbia Association of Mathematics Teachers' Annual Fall Conference. Available at: http://youtube.be/FILJ6FoDLXg

Sinclair, N., & Armstrong, A. (2011). Tell a piecewise story. *Mathematics Teaching in the Middle School, 16*(6), 346–353.

Sinclair, N., & Bruce, C. D. (coordinators). (2014). Research forum: Spatial reasoning for young learners. In P. Liljedahl, C. Nicol, S. Oesterle, & D. Allan (Eds.), *Proceedings of the Joint Meeting of PME 38 and PME-NA 36, vol. 1* (pp. 173–203). Vancouver, BC: PME.

Sinclair, N., & de Freitas, E. (in press). The haptic nature of gesture: rethinking gesture with new multitouch digital technologies. *Gesture*.

Sinclair, N., de Freitas, E., & Ferrara, F. (2013). Virtual encounters: the murky and furtive world of mathematical inventiveness. *ZDM: The International Journal on Mathematics Education, 45*(2), 239–252.

Sinclair, N., & Gol Tabaghi, S. (2010). Drawing space: mathematicians' kinetic conceptions of eigenvectors. *Education Studies in Mathematics, 74*(3), 223–240.

Sinclair, N. , & Moss, J. (2012). The more it changes, the more it becomes the same:

the development of the routine of shape identification in dynamic geometry environments. *International Journal of Education Research, 51&52*, 28–44.

Sorby, S. A. (2009). Developing spatial cognitive skills among middle school students. *Cognitive Processing, 10* (Suppl 2). doi: 10.1007/s10339-009-0310-y

Sorby, S. A. (2012). *Developing spatial thinking*. Clifton Park, NY: Delmar, Cengage Learning.

Sorby, S., Casey, B. M., Veurink, N., & Dulaney, A. (2013). The role of spatial training in improving spatial and calculus performance in engineering students. *Learning and Individual Differences, 26*, 20–29.

Spatial Intelligence and Learning Center Tests and Instruments. (2014). Retrieved from http://spatiallearning.org/index/php/testsainstruments

Spencer, S. J., Steele, C. M., & Quinn, D. M. (1999). Stereotype threat and women's math performance. *Journal of Experimental Social Psychology, 35*(1), 4–28.

Stanic, G., & Kilpatrick, J. (Eds.). (2003). *A history of school mathematics*. Reston, VA: National Council of Teachers of Mathematics.

Steenpaß, A., & Steinbring, H. (2013). Young students' subjective interpretations of mathematical diagrams: elements of the theoretical construct "frame-based interpreting competence." *ZDM: The International Journal on Mathematics Education, 46*, 3–14.

Stenning, K. (2002). *Seeing reason: image and language in learning to think*. Oxford: UK: Oxford University Press.

Stevens, R. (2012). The missing bodies of mathematical thinking and learning have been found. *Journal of the Learning Sciences, 21*(2), 337–346.

Stewart, I. (2007). *Why beauty is truth: a history of symmetry*. New York: Basic Books.

Stull, A. T., Barrett, T., & Hegarty, M. (2013). Usability of concrete and virtual models in chemistry instruction. *Computers in Human Behavior, 29*(6), 2546–2556. doi: 10.1016/j.chb.2013.06.012

Stull, A. T., Hegarty, M., Dixon, B., & Stieff, M. (2012). Representational translation with concrete models in organic chemistry. *Cognition and Instruction, 30*(4), 404–434. doi: 10.1080/07370008.2012.719956

Stumpf, H., & Haldimann, M. (1997). Spatial ability and academic success of sixth grade students at international schools. *School Psychology International, 18*(3), 245–259. doi: 10.1177/0143034397183005

Tahta, D. (1980). About geometry. *For the Learning of Mathematics, 1*(1), 2–9.

Tahta, D. (1989). Is there a geometric imperative? *Mathematics Teaching, 129*, 20–29.

Taylor, H. A., & Hutton, A. (2013). Think3d!: training spatial thinking fundamental to STEM Education. *Cognition and Instruction, 31*(4), 434–455.

Terlecki, M., Newcombe, N., & Little, M. (2008). Durable and generalized effects of spatial experience on mental rotation: gender differences in growth patterns. *Applied Cognitive Psychology, 22*, 996–1013.

Thom, J. S. (2011). Nurturing mathematical reasoning. *Teaching Children's Mathematics, 18*(4), 234–243.

Thom, J. S. (2012). *Re-rooting the learning space: minding where children's mathematics grow*. Rotterdam, NL: Sense.

Thom, J., & McGarvey, L. (in press). The act and artifact of drawing(s) in mathematics: observing geometric thinking through children's drawings. *ZDM: The International Journal on Mathematics Education, 15*(3).

Thom, J., & Roth, W.-M. (2011). Radical embodiment and semiotics: toward a theory of mathematics in the flesh. *Educational Studies in Mathematics, 77*, 267–284.

Thompson, J., Nuerk, H. C., Moeller, K., & Cohen Kadosh, R. (2013). The link between mental rotation ability and basic numerical representations. *Acta Psychologica, 144,* 324–331.

Thurston, W. (1995). On proof and progress in mathematics. *For the Learning of Mathematics, 15*(1), 29–37.

Thurston, W. P. (1994). On proof and progress in mathematics. *The American Mathematical Society, 30*(2), pp. 161–177.

Toth, C. D., O'Rourke, J., & Goodman, J. E. (2004). *Handbook of discrete and computational geometry, 2nd edition.* New York: CRC Press.

Towers, J., & Martin, L. C. (2014). Enactivism and the study of collectivity. *ZDM: The International Journal on Mathematics Education, 47*(2). doi: 10.1007/s11858-014-0643-6

Trafton, J. G. & Harrison, A. M. (2011). Embodied spatial cognition. *Topics in Cognitive Science, 3*(4), 686–706.

Tzuriel, D., & Egozi, G. (2010). Gender differences in spatial ability of young children: the effects of training and processing strategies. *Child Development, 81*(5), 1417–1430.

Uttal, D. H. (2000). Seeing the big picture: map use and the development of spatial cognition. *Developmental Science, 3*(3), 747–286.

Uttal, D. H., Meadow, N. G., Tipton, E., Hand, L. L., Alden, A. R., Warren, C., & Newcombe, N. S. (2013). The malleability of spatial skills: a meta-analysis of training studies. *Psychological Bulletin, 139*(2), 352–402. doi: 10.1037/a0028446

Varela, F. J., Thompson, E., & Rosch, E. (1991). *The embodied mind: cognitive science and human experience.* Cambridge, MA: MIT Press.

Vasilyeva, M., & Bowers, E. (2006). Children's use of geometric information in mapping tasks. *Journal of Experimental Child Psychology, 95*(4), 255–277.

Vasilyeva, M., & Huttenlocher, J. (2004). Early development of scaling ability. *Developmental Psychology, 40*(5), 682–690.

Vasta, R., Knott, J. A., & Gaze, C. E. (1996). Can spatial training erase the gender differences on the water-level task?. *Psychology of Women Quarterly, 20*(4), 549–567.

Verdine, B. N., Golinkoff, R. M., Hirsh-Pasek, K., & Newcombe, N. S. (2014). Finding the missing piece: blocks, puzzles, and shapes fuel school readiness. *Trends in Neuroscience and Education, 7*(1), 7–13. doi.org/10.1016/j.tine.2014.02.005

Verdine, B. N., Golinkoff, R., Hirsh-Pasek, K., Newcombe, N., Filipowocz, A. T., & Chang, A. (2014). Deconstructing building blocks: preschoolers' spatial assembly performance relates to early mathematics skills. *Child Development, 85*(3), 1062–1076. doi: 10.1111/cdev.12165

Verdine, B. N., Troseth, G. L., Hodapp, R. M., & Dykens, E. M. (2008). Strategies and correlates of jigsaw puzzle and visuospatial performance by persons with prader-willi syndrome. *American Journal of Mental Retardation, 113*(5), 343–355.

Voyer, D. (2011). Time limits and gender differences on paper-and-pencil tests of mental rotation: a meta-analysis. *Psychonomic Bulletin & Review, 18*(2), 267–277.

Vygotsky, L. S., & Luria, A. (1993). *Studies on the history of behavior, ape, primitive, and child.* Hillsdale, NJ: Lawrence Erlbaum Associates.

Wai, J., Lubinski, D., & Benbow, C. P. (2009). Spatial ability for STEM domains: aligning over 50 years of cumulative psychological knowledge solidifies its importance. *Journal of Educational Psychology, 101*(4), 817–835.

Watters, E. (2013). *We aren't the world.* Retrieved from http://www.psmag.com/magazines/magazine-feature-story-magazines/joe-henrich-weird-ultimatum-game-shaking-up-psychology-economics-53135/

Werner, K., & Raab, M. (2014). Moving your eyes to solution: effects of movements on the perception of a problem-solving task. *The Quarterly Journal of Experimental Psychology, 67*(8), 1571–1578.

Wheatley, G. (1996). *Quick draw: Developing spatial sense in mathematics.* Tallahassee: Florida Department of Education.

Whiteley, W. (1999). *The decline and rise of geometry in 20th-century North America.* Proceedings of the Annual Meeting of the Canadian Mathematics Education Study Group. http://www.math.yorku.ca/~whiteley/cmesg.pdf

Whiteley, W. (2005). Learning to see like a mathematician. In G. Malcolm (Ed.), *Multidisciplinary approaches to visual representation and interpretation* (pp. 279–292). New York: Elsevier.

Whiteley, W., & Mamolo, A. (2014). Optimizing through geometric reasoning supported by 3D models: visual representations of change. In A. Watson & M. Ohtani (Eds.), *ICMI Study 22 on Task Design* (pp. 129–140). Berlin: Springer.

Whiteley, W., Sinclair, M., Craven, S., Moshe, L., Dutfield, A., & Seco, M. (2008). *Helping elementary teachers develop visual and spatial skills for teaching geometry.* Paper presented at the 11th International Congress on Mathematics Education (ICME 11), Monterrey, MX, July.

Wikipedia. (2014a). *Multiview orthographic projection.* Retrieved August 31, 2014, from http://en.wikipedia.org/wiki/Multiview_orthographic_projection

Wikipedia. (2014b). *Chirality in Chemistry.* Retrieved October 31, 2014, from http://en.wikipedia.org/wiki/Chirality_(chemistry)

Wikipedia. (2014c). *Necker Cube.* Retrieved October 31, 2014, from http://en.wikipedia.org/wiki/Necker_cube

Witkin, H. A. (1971). *A manual for the embedded figures tests.* Palo Alto, CA: Consulting Psychologists Press.

Witkin, H. A., & Asch, S. E. (1948). Studies in space orientation: IV. Further experiments on perception of the upright with displaced visual fields. *Journal of Experimental Psychology, 38,* 762–782.

Witt, M. (2011). School based working memory training: preliminary finding of improvement in children's mathematical performance. *Advances in Cognitive Psychology, 7*(1), 7–15. doi: 10.2478/v10053-008-0083-3

Wolfgang, C. H., Stannard, L. L., & Jones, I. (2001). Block play performance among preschoolers as a predictor of later school achievement in mathematics. *Journal of Research in Childhood Education, 15*(2), 173–180.

Wright, R., Thompson, W., Ganis, G., Newcombe, N., & Kosslyn, S. (2008). Training generalized spatial skills. *Psychonomic Bulletin and Review, 15*(4), 763–771.

Zimmermann, W. (1991). Visual thinking in calculus. In W. Zimmermann & S. Cunningham (Eds.), *Visualization in teaching and learning mathematics* (pp. 127–137). Washington, DC: Mathematical Association of America.

Name Index

Subject Index